Algal Bioprocess Technology

Algal Bioprocess Technology

S. S. Lele
Prof. of Biochemic Engineering
Coordinator Bioprocess Technology
Food Engg. & Tech. Deptt.
U.I.C.T., Mumbai.

Jyoti Kishen Kumar
Scientist 'A'
For Bioprocess Technology
Food Engg. & Tech. Deptt.
U.I.C.T., Mumbai.

PUBLISHING FOR ONE WORLD

NEW AGE INTERNATIONAL (P) LIMITED, PUBLISHERS
(formerly Wiley Eastern Limited)
New Delhi • Bangalore • Chennai • Cochin • Guwahati • Hyderabad
Jalandhar • Kolkata • Lucknow • Mumbai • Ranchi
Visit us at **www.newagepublishers.com**

Published by New Age International (P) Ltd., Publishers
First Edition: 2008

ISBN (10): 81-224-2217-9

ISBN (13): 978-81-224-2217-7

Rs. 200.00

C-07-10-2061

Printed in India at Ramprintograph, Delhi.
Typeset at Inhouse.

PUBLISHING FOR ONE WORLD

NEW AGE INTERNATIONAL (P) LIMITED, PUBLISHERS
(formerly Wiley Eastern Limited)
4835/24, Ansari Road, Daryaganj, New Delhi - 110002
Visit us at **www.newagepublishers.com**

Dedicated

To

Our Beloved Almamater-

UICT

A Temple of Research

Where we learnt

'Work is Worship'

PREFACE

With the phenomenal increase in the world population and industrial growth in the past few decades, some of the major problems confronting mankind are those of quantity and quality of food, feed, drinking water, disposal of sewage and industrial wastes and proper conservation of water and soil. Suitably controlled algal growth can provide help in tackling these problems. Algae have found applications in many areas including agriculture, aquaculture, environmental protection and management and as functional foods and pharmaceuticals.

Bioprocess Technology is an interdisciplinary field encompassing several disciplines such as Microbiology, Biochemistry and Chemical Engineering. Bioprocess Technology mainly comprises of two parts, upstream processing and downstream processing. The upstream processing is mainly related to prefermentation steps and the production (fermentation or biotransformation or enzymatic reaction). Downstream processing is carried out for isolation and purification of the product. Typically, 50–70% of the total production cost in classical processes is due to downstream processing. Downstream processing has been identified as a major bottleneck in commercialization of bioprocesses.

This book, *Algal Bioprocess Technology*, touches upon both upstream and downstream aspects of microalgal cultivation and their potential in commercial applications. The book includes seven chapters. The *first* chapter is intended to give an overview of algae, their diversity, life cycle and reproduction. Major developments in algal biotechnology with respect to the biotechnologically relevant algae have been discussed. The *second* chapter deals with cultivation of algae where, algal nutrition, environmental factors governing their growth and mass cultivation systems are described. Different types of Photobioreactors and their design consideration have been dealt with in detail. *Chapter 3* focuses on high value metabolites from microalgae and various strategies employed in the downstream processing of the same. *Chapter 4* unfolds some interesting studies on biosorption of heavy metals on algal biomass. All aspects of biosorption consisting of its mechanism, influencing parameters, evaluation of sorption performance, immobilized algal biomass, scale up studies in packed columns and novel concepts like use of spent algal biomass are discussed in detail. Keeping in mind the developing area of genetic engineering, yet another important aspect is discussed in *chapter 5,* which unveils the latest achievements in genetic modifications of algae. A few case studies including strategies for transformation of few important microalgal species and its industrial applications are discussed. *Chapter 6* is very unique by itself. Here the reader is not only introduced to the general approaches for mathematical modelling but also exposed to various models for microalgal cultivation, growth kinetics and process development. Some case studies on biosorption equilibrium and kinetic models and recent developments in photobioreactors are also discussed. The book would be incomplete without numerical problems. Hence, these are included in *Chapter 7.* We have tried our best to make the mathematical calculations simple and interesting by incorporating a lot of problems right from cultivation of microalgae to downstream processing. Some parts of this book will reflect our own ideas and personal experiences, which are presented as reviews.

The book has been prepared, both for the specialist and the beginner. It will be useful to postgraduate and undergraduate students of biochemical engineering, biotechnology, industrial biotechnology, bioprocess technology and allied courses. Researchers in this field will also find it worth having this book in their collection.

Last but not least we thank those students of Bioprocess Technology and Food Engineering and Technology Department who enthusiastically participated in collecting material and supported in this endeavor. Special acknowledgement of Sagar Gokhale who single handedly contributed for the solved problems, which has made a significant value addition to this 'product'. We thankfully acknowledge the constant support and encouragement from our colleagues and Director.

We wish to express our thanks to New Age International (P) Limited Publishers and the editorial department for the inspiration given by them to undertake and complete this project. We would welcome comments and suggestions for further improvement of the book.

S.S. Lele
Jyoti Kishen Kumar

CONTENTS

Chapter 1

INTRODUCTION

1.1 ALGAL ORIGIN AND DIVERSITY

For millennia, aquatic environment has been a dwelling place for many simple life forms to complex biological forms of higher order. Algae are one such aquatic forms which have vast resources of biochemicals that have not yet been explored properly. They are a diverse group of organisms some time ago thought to fit into a single class of plants. In the beginning, algae were considered to be simple plants lacking leaf, stem, root and reproductive systems of Higher Plants such as mosses, ferns, conifers and flowering plants. However, it was realized that some of them have animal like characteristics so they were incorporated in both the plant and animal kingdoms. Thus, algae are considered as oxygen producing, photosynthetic organisms that include macroalgae, mainly seaweeds and a diverse group of microorganisms known as microalgae. This book focuses mainly on microalgae. They are photosynthetic and can absorb the sun's energy to digest water and CO_2, releasing the precious atmospheric oxygen that allows the entire food chain to sprout and flourish in all its rich diversity.

Microalgae have many special features, which make them an interesting class of organisms. Many freshwater algae are microscopic in nature. They vary in size ranging from a smallest cell diameter of 1000 mm to largest algal seaweed of 60 m in height. Microalgae are very colourful. They exhibit different colours such as green, brown and red. In general, microalgae have shade between and mixtures of these colors. Most of them can make their own food materials through photosynthesis by using sunlight, water and carbon dioxide. A few of them are not photoautotropic, but they belong to groups, which are usually autotrops. They may be found as free-floating phytoplankton, which form the base of food webs in large water bodies. They can also be found on land attached to various surfaces like steps, roofs etc. There are microalgae, which live, attached to rocks or paving stones and other substrata at the bottom of the sea. They may occur as epiphytes on higher plants, or on other algae. All major bodies of water have these organisms in abundance, including, permanent or semi-permanent water of lakes, small streams, large rivers, reservoirs, ponds, canals and even waterfalls. Most of these organisms can tolerate different degrees of salinity. Some of them dwell in fresh water or sea water whereas some are able to tolerate the extreme salinity of saltpans. In the sea they may occur below the range of tidal exposure — in the sub tidal zone as well as in the harsh intertidal environment of the seashore where they may be beaten by waves. Growing in the intertidal zone, microalgae are subjected to a number of stresses and disturbances. At low tide, they may bake in the sweltering sun or even get rained on by fresh water. In some parts of the world, intertidal microalgae are even scoured by sea ice, yet they persist in living in this environment at 4°C, some even close to freezing point. Those algae, which live attached to the bottom of a water body, are called benthic algae, and the ecosystems of which they are a part are referred to as benthos. The upper limit for their survival is 30°C but there are also algae that thrive at 60°C in the heated water of hot springs. In deserts they are found least common in wind blown sandy deserts and most common in the pebbly, rocky or clayey deserts (Lund, H.C., 1995). Small, microscopic algae, which drift about in bodies of water, such as lakes and oceans, are called phytoplankton. Phytoplanktons are important in freshwater and marine food webs, and are probably responsible for producing much of the oxygen that we breathe. Some forms of algae are able to grow in Arctic and Antarctic sea ice, where they can be quite productive and support a whole associated food web. Some algae can grow on the seabed, beneath a thick blanket of Arctic or Antarctic sea ice, even though they are in total darkness for a considerable part of the year. Algae are found in snow too! In some parts of the world, blooms of snow algae may paint the snow beds red in spring. One may be astonished to find that algae even occur in the driest deserts. In some areas of the Namib Desert in Namibia, and the Richtersveld in South Africa, one often finds many quartz stones scattered about on the ground. Since Quartz is quite translucent, the stones permit a considerable amount of light to pass through, so there is sufficient light for photosynthesis to take place underneath the stones. A small amount of moisture may be retained in the soil under the quartz stones; so unicellular algae are able to grow underneath them. It is amazing to note that algae are also found in the air, for there are many algae that colonize new bodies of water by simply drifting about through the air. Some algae are known to cause diseases in humans. *Prototheca,* a unicellular green alga produces skin lesions, mainly in patients whose immune systems have been damaged by other serious diseases.

Some species of algae form symbiotic relationships with other organisms. In these symbioses, the algae supply photosynthates (organic substances) to the host organism providing protection to the algal cells. The host organism derives some or all of its energy requirements from the algae. Examples include:

- *Lichens* – a fungus is the host, usually with a green alga or a cyanobacterium as its symbiont. Both fungal and algal species found in lichens are capable of living independently, although habitat requirements may be greatly different from those of the lichen pair.
- *Corals* – algae known as zooxanthellae are symbionts with corals. Notable amongst these is the dinoflagellate *Symbiodinium*, found in many hard corals. The loss of *Symbiodinium*, or other zooxanthellae, from the host is known as coral bleaching.

- *Sponges*—green algae live close to the surface of some sponges, for example, breadcrumb sponge (*Halichondria panicea*). The alga is thus protected from predators; the sponge is provided with oxygen and sugars which can account for 50 to 80% of sponge growth in some species.

This fascinating group of organisms forms the basis for the science of **Phycology— the study of algae.**

1.2 CLASSIFICATION

To date, algae have been classified in terms of various parameters like pigments, flagella, reserve material, habitats, size, shape and cell wall composition. A detailed classification of algae is presented in Table 1.1 and Table 1.2. Organisms that make up the algae include representatives from three kingdoms and seven divisions: Cyanochloranta and Prochorophyta (from Kingdom Monera), Pyrrhophyta, Chrysophyta, Phaeophyta, and Rhodophyta (from Kingdom Protista), and Chlorophyta (from Kingdom Plantae). All seven divisions are called algae because of a lack of roots stems and leaves; and most algal cells are fertile. The basic metabolic processes are located in the individual cell and all lack the xylem/phloem transport system of "higher plants". These different plant-like organisms have been used for human food and animal follage.

Table 1.1 **Classification based on characteristics and habitat**

Algal Class	Example	Characteristics	Habitat
Cyanophyta	*Synechocystis, Spirulina*	Bluegreen, Buoyant, Gliding	Lakes, Streams
Chlorophyta	*Chlamydomonas, Dunaliella, Haematococcus*	Green, Flagellated	Freshwater, Lakes, Rivers
Euglenophyta	*Euglena*	Varied in colour, Flagellated	Lakes, Ponds
Eustigmatophyta, Raphidiophyta, Tribophyta	*Vischeria*	Yellow green, Flagellated and Nonflagellated	Benthic, Epiphytic
Dinophyta	*Ceratium*	Reddish Brown, Flagellated	Lakes, Estuaries
Cryptophyta	*Rhodomonas, Cryptomonas*	Varied in colour Flagellated	Lakes, Planktonic
Chryophyta	*Mallomonas, Dinobryon*	Golden, Flagellated	Lakes, Streams

Contd...

Chryophyta	*Mallomonas, Dinobryon*	Golden, Flagellated	Lakes, Streams
Bacillariophyta	*Stephanodiscus, Aulacoseira*	Golden Brown, Gliding	Lakes, Estuaries, Planktonic
Rhodophyta	Batrachospermum	Red, Nonmotile	Streams, Lakes
Phaeophyta	*Pleurocladia, Heribaudiella*	Brown, Nonmotile	Streams, Lakes

Table 1.2 Classification based on cell wall composition and reserve material

Algal class	Cell wall composition	Reserve material
Cyanophyta	Peptidoglycan	Cyanophycean starch
Chlorophyta	Cellulose	True starch
Euglenophyta	Protein	Paramylon
Eustigmatophyta, Raphidiophyta, Tribophyta	Cellulose	Chrysolaminarin
Dinophyta	Cellulose or no cell wall	True starch
Cryptophyta	Cellulose periplast	True starch
Chrysophyta	Pectin, Silica	Chrysolaminarin
Bacillariophyta	Silica fustules	Chrysolaminarin
Rhodophyta	Galactose polymer	Floridean starch
Phaeophyta	Alginate	Laminarin

1.3 LIFE CYCLE AND REPRODUCTION

A spectacular diversity is seen in algal reproduction. Asexual reproduction is observed in some algae while sexual reproduction is noticed in a few; others follow both of these mechanisms for multiplication. Asexual reproduction is accomplished by binary fission where an individual cell breaks into two, which is often seen, in unicellular algal members. Most algae are capable of reproducing by spores, these spores on dissemination from the parent alga grows into new individuals under favorable conditions. Sexual reproduction however is restricted to multi-cellular forms where the union of cells takes place through a process called conjugation. As case studies, the life cycle history of blue green alga (*Spirulina platensis*) and the chlorphyte (*Haematococcus pluvialis*) are discussed below:

Spirulina plantensis

The life cycle of Spirulina is relatively simple. The trichome on maturing breaks into many fragments by forming special cells called necridia. These necridia undergo lysis to form biconcave separation disks. Thereafter, fragmentation of trichome at necridia results in a short gliding chain of harmogonia. These specialized cells (harmogonia) detach from the parent filament and give rise to new trichome. The cells found in hormogonium lose the necridia cells and become round at the distal ends with very little thickening of the cell wall. In due course of this process the cell cytoplasm appear less granulated and the cells turn pale blue-green in color. The cells in harmogonia increase by cell fission and the cell cytoplasm now becomes granulated. The cell assumes bright blue green color. This process results in trichomes, which grow by length and turn into the typical helical shape. The spontaneous breakage of trichomes with formation of necridia is rarely seen in this organism. Akinetes (reproductive spores) is however not been reported in this organism.

Haematococcus pluvialis

Haematococcus pluvialis, a green chlorphyte is a flagellated unicellular microalga. This alga is known to accumulate large amount red pigment astaxanthin that is produced during encystment stage during adverse environmental conditions like light intensity, nutritional deficiency etc. During its life cycle four types of cells were distinguished: microzooids, large flagellated macro-zooids, non-motile palmella forms; and haematocysts, which are large red cells with a heavy resistant wall. The macro-zooids is generally the most predominated form found in liquid cultures with sufficient nutrients. However during extreme unfavorable environmental conditions, the palmella stage changes to haematocysts, which accumulate red colored astaxanthin. Haematocysts however when exposed to favorable conditions (nutrients or environmental conditions) gives rise to motile micro-zooids that either grow into palmella or macro-zooid stages.

1.4 BIOTECHNOLOGICALLY RELEVANT MICROALGAE

The color green has been associated with healing throughout history, spanning continents and many religions. Green also signifies new life, growth and regeneration. The explosive nutritive value found in a microscopic algae equivalent to the size a single human blood cell is what makes them 'super foods', packing big supplemental punch. The 'macroalgae', usually referred to as seaweed, have been commercially cultured for over 300 years (Tseng, 1981). Most people in the United States ingest red or brown algal products everyday in chocolate milk, toothpaste, candy, cosmetics, ice creams, salad dressing, and many other household and industrial products (McCoy, 1987). Macroalgae are rich in protein, carbohydrates, amino acids, trace elements, and vitamins (Waaland, 1981). Historically, records have established that people collected seaweeds for food beginning 2,500 years ago in China (Tseng, 1981). European people have collected seaweeds for food for 500 years. Today, only in the Far East are macroalgae eaten directly in large quantities as food by humans. The typical Porphyrrean algae are called 'Nori', 'amanori' or 'hoshinori' in Japan and 'purple laver' in the West. This genus of red algae represents the largest tonnage aquacultural product in the world (McCoy, 1987) and was the first marine macroalgae to be cultivated by man. Nori has been grown in Tokyo Bay for nearly 300 years (Lobban et al., 1985). Nori is eaten directly in soups, as a vegetable or used as a condiment. The Japanese grow over 500,000 tons of Nori per year and consume over 100,000 tons directly per year. The Nori industry in Japan employs over 60,000 people and is estimated to support over 300,000 people (McCoy, 1987). The Chinese also have a very large Nori industry but no estimation on the number of employees has been given. Major commercial centers for Nori include Marinan Islands, Saipan, and Guam. However, the world's largest and most technically advanced Nori farms are facilities in the Philippines (McCoy, 1987). Nori is also eaten in Europe, mainly in salads, fried in fat, boiled and even baked into bread. The British used to seal the fresh algae in barrels for use as food by whaling crews. In the United States, Nori is commonly found in health food stores. Nori is also used in the preparation of 'sushi'. The alga is wrapped around the raw seafood and rice to hold the two together. The majority of the macroalgae that is under cultivation are used for their phycocolloids. There are three major commercial groups of phycocolloids: agar-agar, algins and caregeenans. The estimated world market value for phycocolloids is US $550 million (www.siu.edu).

The primary agar producing genera are *Euchema, Gelidium, Gracilaria, Hypnea, Gigartina* and *Marocystis* (Chapman, 1970). The name agar comes from the native Malaysian name for Euchema, 'agar-agar' (Tseng, 1981). Agar is a group of complex entities made up of calcium or magnesium salts of a sulfuric acid ester of a linear galactan. This substance is a major constituent of the cell wall of some red algae. Agar has been used extensively in microbiology for culturing instead of gelatins because of its ability to remain a semi-solid at 0°C to 70°C, it has a low viscosity when melted, ease of mixing and pouring, firmness and clarity of agar gels. Unlike gelatins, most species of bacteria cannot digest agar. With the advent of modern molecular biology and genetic engineering, agar gums producing an 'agrarose' factor is used extensively in electrophoresis and chromatography. Agrarose gel electrophoresis has replaced starch gel electrophoresis in most laboratories around the world.

Carrageenan is a phycocolloid much like agar. This compound is a family of sulfated galactan polymers obtained from various red algae especially *Chondrus, Sigartina, Iridaea, Hypnea,* and *Eucheuma.* Originally, carrageean was processed from Irish moss, *Chondrus crispis.* Carrageenans are generally employed for their physical functions in gelation, viscous behavior, stabilization of emulsions, suspensions, foams and control of crystal growth (Chapman, 1970). Other applications of carrageenans include uses in pharmaceutical, cosmetics and various coatings such as paints and inks. It is also commonly used in items like ice cream and pudding.

The third class of phycolloids is the algins or algenic acids. Algins are a major constituent of all brown algae. Chemically it is a polymer of d-mannuronic and I-guluronic acids. There are some 897 known chemical members of this family. Alginic acids are commercially important in the production of rubber and textiles. Before World War II, Japan was the only major producer of algenic acid. During the war, California algenic acid industry was made. The salts of algins produce a clear, tough film, which is used extensively as thickeners, coagulants, or flocculants in many foods. Examples include soups, mayonnaise, sauces and sausage casings (McCoy, 1987). There are several species of brown algae harvested currently; most commercially important algins come from the giant kelp, *Macrocystis* and *Nerocystis.* Asian societies have used algae for centuries as a source of folk medicine, soil conditioners and food. The low-density, labor-intensive farming of edible seaweeds such as Nori (Porphyra) off the Japanese and Korean coasts constitutes a $ 1.5 billion a year industry. In the western countries natural populations of seaweeds are principally harvested for their gel content, which is processed into agar and carageenan for industrial and food thickeners and biological culture media. Algae represent a major bio-resource today. Of the 150,000 species estimated to exist, more than 30,000 have been identified. Yet the basic taxonomy of many algal species is incomplete. Developing algae for commercial use depends on selecting, screening and culturing natural species due to which advances in mass culture technology mainly aimed at manipulating environmental conditions to enhance quality and quantity of the alga had largest impact. Some microalgae with biotechnological relevance are discussed.

Brown algae

The Phaeophyta or the brown algae are a large group of multicellular, mostly marine, algae, including many notable seaweeds of northern waters. They play an important role in marine environments. For instance *Macrocystis,* a member of the Laminariales or kelps, may reach 60 metres in length and forms prominent underwater forests. Another notable example is *Sargassum,* which creates unique habitats in the Sargasso Sea (hence the name *Sargassum*). Many brown algae such as members of the order Fucales (the rockweeds) are commonly found along rocky seashores. Some members of the division are used as food. Brown algae belong to a very large group called the heterokonts, most of which are colored flagellates. Most contain the pigment fucoxanthin, which is responsible for the distinctive greenish-brown color that gives brown algae their name. Brown algae are unique among heterokonts in developing into multicellular forms with differentiated tissues, but they reproduce by means of flagellate spores, which closely resemble other heterokont cells. Genetic studies show their closest relatives are the yellow-green algae.

Green algae

The Chlorophyta, or green algae, include about 17,000 species of mostly aquatic photosynthetic eukaryotic organisms. Like the land plants (Bryophyta and Tracheophyta), green algae contain chlorophylls *a* and *b,* and store food as starch in their plastids. They are related to the Charophyta and Embryophyta (land plants), together making up the Viridiplantae. They contain both unicellular and multicellular species. While most species live in freshwater habitats and a large number in marine habitats, other species are adapted to a wide range of environments. Watermelon snow, or *Chlamydomonas nivalis,* of the class Chlorophyceae, lives on summer alpine snowfields. Others live attached to rocks or woody parts of trees. Some lichens are symbiotic relationships with fungi and a green alga. Members of the Chlorophyta also form symbiotic relationships with protozoa, sponges and coelenterates.

Golden algae

The golden algae or chrysophytes are a large group of heterokont algae, found mostly in freshwater. Originally they were taken to include all such forms except the diatoms and multicellular brown algae, but since then they have been divided into several different groups based on pigmentation and cell structure. They are now usually restricted to a core group of closely related forms, distinguished primarily by the structure of the flagella in motile cells, also treated as an order Chromulinales. They come in a variety of morphological types, originally treated as separate orders or families. Most members are unicellular flagellates, with either two visible flagella, as in *Ochromonas,* or sometimes one, as in *Chromulina*. The Chromulinales included only the latter type, with the former treated as the order Ochromonadales. However, structural studies have revealed that short second flagellum or at least a second basal body is always present, so this is no longer considered a valid distinction. Most of these have no cell covering. Some have loricae or shells, such as *Dinobryon,* which is sessile and grows in branched colonies. Most forms with silicaceous scales are now considered a separate group, the synurids, but a few belong among the Chromulinales proper, such as *Paraphysomonas.* Some members are generally amoeboid, with long branching cell extensions, though they pass through flagellate stages as well. *Chrysamoeba* and *Rhizochrysis* are typical of these. There is also one species, *Myxochrysis paradoxa,* which has a complex life cycle involving a multinucleate plasmodial stage, similar to those found in slime moulds. These were originally treated as the order Chrysamoebales. The superficially similar *Rhizochromulina* was once included here, but is now given its own order based on differences in the structure of the flagellate stage. Other members are non-motile. Cells may be naked and empeded in mucilage, such as *Chrysosaccus,* or coccoid and surrounded by a cell wall, as in *Chrysosphaera.* A few are filamentous or even parenchymatous in organization, such as *Phaeoplaca.* These were included in various older orders, most of the members of which are now included in separate groups. *Hydrurus* and its allies, freshwater genera which form branched gelatinous filaments, are often placed in the separate order Hydrurales but may belong here.

Red algae

The red algae, *Rhodophyta,* are a large group of mostly multicellular, marine algae, including many notable seaweeds. Most of the coralline algae, which secrete calcium carbonate and play a major role in building coral reefs, belong here. Red algae such as dulse and nori are a traditional part of European and Asian cuisine and are used to make other products like agar, carrageenans and other food additives. Many red algae have multicellular stages but these lack differentiated tissues and organs. Unlike most other algae, no cells with a flagellum are found in any member of the group. Unicellular forms typically live attached to surfaces rather than floating among the plankton, and both the larger female and smaller male gametes are non-motile, so that most have a low chance of fertilization. They have cell walls that are made out of cellulose and thick gelatinous polysaccharides which are the basis for most of the industrial products made from red algae. The chloroplasts of red algae are bound by a double membrane, like those of green plants; both groups (*Archaeplastida*) probably share a common origin. Their plastids formed by direct endosymbiosis of a cyanobacteria, and in red algae are pigmented with chlorophyll *a* and various proteins called phycobiliproteins, which are responsible for their reddish color. Other reddish algae are classified not as red algae but as Chromista which are hypothesied to have acquired their chloroplasts from red algae through endosymbiosis. The oldest fossil identified as a red alga is also the oldest fossil eukaryote that belongs to a specific modern taxon. *Bangiomorpha pubescens,* a multicellular fossil from arctic Canada, strongly resembles the modern red alga *Bangia* despite occurring in rocks dating to 1200 million years ago. Red algae are important builders of limestone reefs. The earliest such coralline algae, the solenopores, are known from the Cambrian Period. Other algae of different origins filled a similar role in the late Paleozoic, and in more recent reefs.

Blue-green algae

Cyanobacteria are often referred to as blue-green algae. They obtain their energy through photosynthesis. The description is primarily used to reflect their appearance and ecological role rather than their evolutionary lineage. Fossil traces of cyanobacteria have been found from around 3.8 billion years ago. As soon as these blue-green bacteria evolved, they became the dominant metabolism for producing fixed carbon in the form of sugars from carbon dioxide. Cyanobacteria are now one of the largest and most important groups of bacteria on earth. The cyanobacteria were traditionally classified by morphology into five sections, referred to by the numerals I-V. The first three—*Chroococcales, Pleurocapsales* and *Oscillatoriales* — are not supported by phylogenetic studies. However, the latter two—Nostocales and Stigonematales—are monophyletic and make up the heterocystous cyanobacteria. Cyanobacteria are found in almost every conceivable habitat, from oceans to fresh water to bare rock to soil. They may be single-celled or colonial. Colonies may form filaments, sheets or even hollow balls. Cyanobacteria include unicellular, colonial and filamentous forms. Some filamentous colonies show the ability to differentiate into three different cell types: vegetative cells are the normal, photosynthetic cells that are formed under favorable growing conditions; akinetes are the climate-resistant spores that may form when environmental conditions become harsh; and thick-walled heterocysts that contain the enzyme nitrogenase, vital for nitrogen fixation, that may also form under the appropriate environmental conditions wherever nitrogen is present. Carbon dioxide is reduced to form carbohydrates via the Calvin cycle. Due to their ability to fix nitrogen in aerobic conditions they are often found as symbionts with a number of other groups of organisms such as fungi (lichens), corals, pteridophytes (Azolla), angiosperms (Gunnera) etc. Cyanobacteria are the only group of organisms that are able to reduce nitrogen and carbon in aerobic conditions, a fact that may be responsible for their evolutionary and ecological success. The water-oxidizing photosynthesis is accomplished by coupling the activity of photo system (PS) II and I. They are also able to use in anaerobic conditions only PS I — cyclic photophosphorylation —with electron donors other than water (hydrogen sulfide, thiosulphate, or even molecular hydrogen) just like purple photosynthetic bacteria. Chlorophyll *a* and several accessory pigments (phycoerythrin and phycocyanin) are embedded in photosynthetic lamellae, the analogs of the eukaryotic thylakoid membranes. The photosynthetic pigments impart a rainbow of possible colors: yellow, red, violet, green, deep blue and blue-green cyanobacteria are known. A few genera, however, lack phycobilins and have chlorophyll *b* as well as chlorophyll *a*, giving them a bright green colour.

Diatoms

Diatoms are a major group of eukaryotic algae and are one of the most common types of phytoplankton. Most diatoms are unicellular, although some form chains or simple colonies. A characteristic feature of diatom cells is that they are encased within a unique cell wall made of silica. These walls show a wide diversity in form, some quite beautiful and ornate, but usually consist of two symmetrical sides with a split between them, hence the group name. Diatoms are a widespread group and can be found in the oceans, in freshwater, in soils and on damp surfaces. Most live pelagically in open water, although some live as surface films at the water-sediment interface (benthic), or even under damp atmospheric conditions. They are especially important in oceans, where they are estimated to contribute up to 45% of the total oceanic primary production. Diatoms belong to a large group called the heterokonts, including both autotrophs (e.g. golden algae, kelp) and heterotrophs (e.g. water moulds). Their chloroplasts are typical of heterokonts, with four membranes and containing pigments such as fucoxanthin. Individuals usually lack flagella, but they are present in gametes and have the usual heterokont structure, except they lack the hairs (mastigonemes) characteristic in other groups. Most diatom species are non-motile but some are capable of an oozing motion. As their relatively dense cell walls cause them to readily sink, planktonic forms in open water usually rely on turbulent mixing of the upper layers by the wind to keep them suspended in sunlit surface waters. Some species actively regulate their buoyancy to counter sinking. Diatoms cells are contained within a unique silicate (silicic acid) cell wall comprised of two separate valves (or shells). The biogenic silica that the cell wall is composed of is synthesised intracellularly by the polymerisation of silicic acid monomers. This material is then extruded to the cell exterior and added to the wall. Diatom cell walls are also called frustules or tests, and their two valves typically overlap one other like the two halves of a petri dish. In most species, when a diatom divides to produce two daughter cells, each cell keeps one of the two valves and grows a smaller valve within it. As a result, after each division cycle the average size of diatom cells in the population gets smaller.

1.5 CURRENT SCENARIO

Many laboratories worldwide are actively involved in perfecting the technology of algal cultivation for various purposes. Nutraceuticals is an umbrella term for dietary supplements containing vitamins and minerals, functional foods, alternative-therapeutic foods. Search for dietary supplements formulated for people with specific diseases, herbal tonics and supplements to tackle vigor and vitality issues has given birth to this new field of "Functional Foods" and "Nutraceuticals". Thus the tenet, "let food be the medicine and medicine be the food", espoused by Hippocrates nearly 2500 ago is receiving renewed interests. The US Institute of Medicines Foods and Nutritional Board defined functional foods as "any food or food ingredient that may provide a health benefit beyond the traditional nutrient it contains". The global nutraceuticals market grew to $ 46.7 billion in 2002, at an Annual Average Growth Rate (AAGR) of nearly 7%. In 2007 nutraceutical sales are projected to reach $ 74.7 billion at an AAGR of 9.9% (www.bccresearch.com). The journey of nutraceuticals as alternative health-care agents is progressing from nutritional supplements to anti-obesity agents to antibiotics to immunomodulatory and anti-carcinogenic agents in piecemeal. Cyanobacteria and microalgae offer a variety of colored compounds like carotenoids, phycobiliproteins and chlorophyll (Borowitzka, 1992). Among these, astaxanthin, β-carotene and phycocyanin have achieved a significant commercial success. Demand of chemically synthesized colorants having greater environmental and human health hazard is decreasing and day by day, there is an increase in the tendency of the consumer to opt for natural and safer colorants of biological origin. This shift in the picture towards search of better alternatives and 'natural' products is welcomed since many of the natural pigments are known to have nutraceutical effect. With abundant solar energy, India has an excellent potential to be a microalgal grower. In fact, India is one of the few countries producing *Spirulina* on commercial scale. Now, there is a need to focus on process development of other microalgae as a whole as well as for nutraceutics and value added products. Commercial scale cultivation of algal culture needs good sunlight, appropriate climatic conditions and desired environment. Most of the advanced countries lack this climatic condition. As a result, there are very few groups of researchers who have explored the area of microalgae and algal-based value added products. Further, extremely slow growth rates of microalgae compared to other microorganisms have resulted in the fact that it is a less explored area for applied research. Major work is done in Japan, Israel, United States and Australia with a few patented processes. Betatene Ltd., Australia; Cyanotech Corporation, USA; Mera Pharmaceuticals, USA; Nature Beta Technologies (NBT), Japan are actively involved in carotenoid production from microalgae (Dufosse et al., 2005). On the local scenario, a few Indian companies are exploiting the virgin market of nutraceuticals that has a projected growth of 30% per annum. Among them are Nicolas Piramal, Ajanta Pharma, Parry Neutraceuticals and Strides Acrolab, to name a few. However, most of the products are herbal and other type, are not based on microalgae. The current focus of these companies is on vitamins, antioxidant and anti-obesity products. But with abundant market available and vast growth rates, the industries are ought to expand to more advanced nutraceuticals e.g. as anticancer and immunomodulatory agents like astaxanthin from microalgae as well microalgae as human food. Dabur has introduced *Spirulina* as a health tonic. The credit of setting up the first commercial plant for *Spirulina* in 1986 goes to the Murrugappa Chettiar Research Centre [MCRC] (Venkataraman, 1995). The Indian Company, Parry's Nutraceuticals is actively involved in commercial microalgal cultivation for high value products (Dufosse et al., 2005). Considering the geographical status, the propitious eight months golden sunlight a year makes India an ideal cultivation field for microalgae; which has been a little exploited industrially. Some of the great challenges to any tropical, marine ornamental and populated industry like India are provision of a consistent, economic and natural health food and pharmaceutical products using available natural resources. The potential of production of natural colorants from microalgae by a techno-economic process needs to be exploited by developing cheaper, energy efficient photobioreactors preferably using solar energy at the larger scale.

1.6 FUTURE TRENDS

A number of commercial developments have occurred in microalgal biotechnology in recent years. New products are being developed for use in the mass commercial markets as opposed to the health food markets. These include algal derived long chained polyunsaturated fatty acids, mainly docosahexanoic acid (DHA) and eicosapentanoic acid (EPA) for use as supplements in human nutrition and animals, pigments in food and pharmaceutical industry, aquaculture and poultry, fertilizers and agrochemicals, for effluent treatment and algae for other bioactive compounds (Borowitzka, 1992). Limiting effects of salt on wastewater treatment are now overcome by replacing conventional sludge by microalga like *Dunaliella* that are well adapted to hypersaline media (Santos et al., 2001). Large scale production of algal fatty acids has been possible due to the use of heterotrophic algae and the adaptation of classical fermentation systems providing consistent biomass under highly controlled conditions resulting in high quality and quantity of products. Algal products have also been developed for use in the pharmaceutical industry. These include stable isotope biochemicals produced by algae in closed systems and extremely bright fluorescent pigments. Some of these potential application areas are discussed further.

High Value Nutraceuticals

Considerable attention has now been directed on the use of algal oils containing long chain polyunsaturated fatty acids (LCPUFAs) as nutritional supplements (Cohen et al., 1995). DHA is the dominant fatty acid in neurological tissue, consisting of 20–25% of the total fatty acid in the gray matter of the human brain and 50–60% in retina rod outer segments (Gill et al., 1997). It is also abundant in heart and muscle tissue and sperm cells. Changes in the EPA levels can change an individual's coronary vascular status as the products of EPA metabolism are eicosanoids with antithrombotic and antiaggregatory effect. Human capacity to produce these oils is poor and hence it has to be supplied in the diet. A number of algal groups have been identified that produce high levels of LCPUFAs, including diatoms, chrysophytes, cryptophytes and dinoflagellates. Dinoflagellates are especially well suited for the production of DHA. The dinoflagellate *Crypthecodinium cohnii* can produce most of its fatty acid as DHA (Behrens et al., 1996). DHA enriched vegetarian oil derived from *Crypthecodinium* is currently widely distributed in the US for the health food market (Brower, 1998). Another DHA enriched product derived from *Schizochytrium* has become available for use as an animal feed. Algae can also provide the genes involved in PUFA synthesis. Once the genes are isolated and characterized their evaluation for suitability for transfer into other organisms and higher plants can be done (Yuan et al., 1997). Algae derived additives are widely used in products like salad dressing, cake frosting, ice-cream, and toothpaste. In addition to the direct use of algae as foods and food supplement algal extracts have potential applications such as preservatives, colorants, vitamins and flavor enhancers (Harvey, 1988).

Aquaculture

Diseases in aquaculture feeds often lead to massive mortality and reduced product quality resulting in heavy financial losses in the fish farmers. It is essential that aquaculture animals obtain their nutrients from the basic algal food chain and the nutrient properties of algae are critical for growth and survival of larvae and adults. In a typical food chain algae are consumed by zooplankton, which in turn are consumed by fish larvae. Improvement of larval nutrition to achieve higher larval survival rate is a challenge for aquaculture industry. Numerous operations that have mastered the art and science of propagation have failed to successfully market their fish as a result of the loss of pigmentation. Given the substantial cost of maintaining the food chain for larvae, any increase in larval survival can have a significant impact of the economics of the aquaculture facility. Nutritional factors have been shown to modulate immune responses in fish. High levels of vitamin C have been reported to increase humoral immunity and serum complement activity (Lygren et al., 1999). Fat-soluble antioxidant vitamin E has been related to increased disease resistance. *Schizochytrium, Crypthecodinium* species containing high quantity of essential oils are used as a source of DHA in fish feed (Barclay et al., 1996). Physiological condition of the fish is a key factor underlying attainment of the required performance level. Algal species commonly cultured for aquaculture feed are *Chlorella, Dunaliella, Tetraselmis, Isohrysis, Navicula, Skeletonema* and *Haematococcus.* These algae are known to be a source of pigmentation to these fishes affecting their commercial acceptability. Carotenoids already are natural constituents of fish-food and help the requirement of fish with better flesh quality and appearance.

Speciality Compounds

One of the speciality compounds from microalgae is fluorescent pigment. Many algal photosynthetic pigments have been well characterized and a number of them are being well utilized for commercial applications. The most widely used are the phycobiliproteins especially in immunodiagnostics and similar assays (Zoha et al., 1998). Phycobiliproteins are a family of light harvesting macromolecules that function as components of the photosynthetic apparatus in Cyanobacteria and several groups of eukaryotic algae like *Cryptomonads* (Apt et al., 1999). The major qualities like having large number of chromophores and high quantum yields, water solubility, forming stable conjugates with many materials, easy excitement by argon or helium-neon lasers makes them most suitable for applications in immunoassays. This allows phycobiliproteins to function as fluorescent tags for labeling highly specific probes to identify cell types or proteins. More significant applications are in flow cytometry and in fluorescence activated cell sorting. Biliproteins have been widely used in immunohistochemistry (Glazer, 1994). Stable Isotopes are another interesting class of compounds that can be obtained from microalgae. Microalgae are ideally suited as the sources of stable isotopically labeled compounds. Their ability to perform photosynthesis allows them to incorporate ^{13}C, ^{15}N and ^{2}H from relatively inexpensive inorganic compounds into more highly valued organic compounds. An example of algal produced stable isotopically labelled complex organic compound is forming the basis of culture media of bacteria, yeast and mammalian cells (Apt et al., 1999). Stable isotopes provided in the media are incorporated into cellular components. Proteins of interest can be produced in large quantity using molecular technology and coupled with recent developments in multidimensional NMR technology and stable isotope editing techniques with structure determination to predict the interaction of substrates with active sites of proteins (Weller et al., 1996). Two commonly used stable isotopically labeled compounds are glucose and glycerol. Microalgae, mainly chlorophytes are known to accumulate large amounts of glucose as starch (Behrens et al., 1989). *Dunaliella* is known to produce high levels of glycerol and has been used for ^{13}C-glycerol production. Several *Chlamydomonas* species are known to produce high level of galactose containing polysaccharide that can be hydrolysed to produce monosacharrides (Behrens et al., 1996). ^{13}C-galactose has been used to measure liver function as its noninvasive nature gives it an advantage over liver biopsy. Similarly, ^{13}C-xylose from *Chlamydomonas* has been used to diagnose bacterial overgrowth of small intestine (Dellert et al., 1997). Microalgae also have the potential to be a rich source of bioactive compounds. A large number of bioactivities including anticancer, antimicrobial, anti-HIV, antiviral and various neurological activities have been reported in algae (Shanbhag, 2001). Certain bluegreen algae and dinoflagellates are also known to be a source of highly potent toxins having significant bioactive effect on humans and fish (Skulberg, 2000). Work at NCI (National Cancer Institute) has demonstrated that sulfoplipids and cyanovirin from microalgae had invitro activity against HIV (Gustafson et al., 1989; Boyd et al., 1997).

Waste Water Treatment

One of the important applications of algae is biosorption of heavy metals. This has been dealt with in Chapter 4. Microalgae can also be used for removal of nutrients, organic contaminants and pathogens from domestic waste water. They play an important role during tertiary treatment of domestic waste water in maturation ponds. They are also used in the treatment of municipal wastewater in facultative or aerobic ponds. During photosynthesis, oxygen is produced which reduces the need of external aeration. This is helpful in the treatment of some hazardous pollutants, which are biodegraded aerobically but may volatilize during mechanical agitation. The following table presents some of the environmental applications of algae.

Table 1.3 Microalgae for wastewater treatment

Application	Comment	References
BOD removal	Microalgae release 1.5–1.92 kg O_2 kg^{-1} of microalgae produced during photoautotrophic growth and oxygenation rates of 0.48–1.85 kg $O_2 m^{-3} d^{-1}$ have been reported in pilot-scale ponds or lab-scale tank photobioreactors treating municipal or artificially contaminated wastewater	Grobbelaar et al., 1988; Martinez Sancho et al., 1993; McGriff and McKinney, 1972; Munöz et al., 2004; Oswald, 1988
Nutrient removal	Microalgae assimilate a significant amount of nutrients because they require high amounts of nitrogen and phosphorous for proteins (45–60% of microalgae dry weight), nucleic acids and phospholipids synthesis. Nutrient removal can also be further increased by NH_3 stripping or P precipitation due to the raise in the pH associated with photosynthesis	Laliberte´ et al., 1994; Oswald, 2003; McGriff and McKinney, 1972; Nurdogan and Oswald, 1995; Vollenweider, 1995

Contd...

Heavy metal removal	Photosynthetic microorganisms can accumulate heavy metals by physical adsorption, ion exchange and chemisorption, covalent bonding, surface precipitation, redox reactions or crystallization on the cell surface. Active uptake that often involves the transport of metals into the cell interior is often a defensive tool to avoid poisoning or it serves to accumulate essential trace elements. Microalgae can also release extracellular metabolites, capable of chelating metal ions. Finally, the increase in pH associated with microalgae growth can enhance heavy metal precipitation	Chojnacka et al., 2005 Kaplan et al., 1995; Kaplan et al., 1987; Rose et al., 1998; Travieso et al., 1996; Van Hille et al., 1999. Wilde and Benemann, 1993; Yu and Wang, which are 2004
Pathogen removal	Microalgae enhance the deactivation of pathogens by raising the pH value, the temperature and the dissolved oxygen concentration of the treated effluent	Aiba, 1982; Mallick,2002 Mezrioui et al., 1994 ; Robinson, 1998; Schumacher et al., 2003
Heterotrophic pollutant removal	Certain green microalgae and cyanobacteria are able to use toxic recalcitrant compounds as carbon, nitrogen, sulphur or phosphorous source	Semple et al., 1999; Subaramaniana and Uma, 1997
Biogas production	CH_4 production from the anaerobic digestion of algal–bacterial biomass allows substantial economical savings	Eisenberg et al., 1981; Oswald, 1988
Toxicity monitoring	Microalgae are used in toxicity tests or in studies of microbial ecology as they are sensitive indicators of ecological changes	Day et al., 1999

Effluents containing organonitriles are highly toxic and sometimes exhibit carcinogenic effects on aquatic life (Nawaz et al., 1989). Organonitriles include acrylonitrile, acetonitrile or cyanide. They are commonly found in effluents from acrylonitrile production plants, polymers or metal plating industries. Physical/chemical treatments conventionally used are alkaline chlorination or oxidation using hydrogen peroxide. But such processes are costly and produce secondary pollution (Augugliaro et al., 1997; Nagle et al., 1995). *Chlorella sorokiniana* in combination with bacterial culture, degrades acetonitrile at a rate of 1.9 and 2.3 g/l in stirred photobioreactor and column photobioreactor (Munõz *et al.,* 2005) with the retention time of 0.6 and 0.4 days respectively. The microbial culture was capable of assimilating upto 71% and nitrifying upto 12% of the NH_4^+ theoretically released from biodegradation of acetonitrile with the retention time of 35 days. N-organics can be completely removed combined with significant removal of nitrogen, using algal-bacterial systems. Aerobic treatment of acetonitriles transforms the pollutants into their corresponding carboxylic acids and ammonia. These carboxylic acids are then further metabolized into CO_2 and H_2O. The problems in this process are high volatility of these compounds, which leads to stripping of the pollutants during process and production of effluents highly loaded with the metabolically produced NH_4^+ that is responsible for the eutrophication of fresh and marine water bodies. In order to reduce NH_4^+ concentration, nitrification and denitrification stages need to be implemented in the treatment process, which increases the cost of the treatment. The use of microalgae could overcome the problems by means of production of O_2^- in photosynthesis process and the ability to assimilate large amount of nutrients. The use of algal-bacterial system allows mitigation of greenhouse effect and at the same time avoids volatilization associated problems due to air sparging. In the mix culture of algae-bacteria, response of photosynthetic micro-organism is species dependant and pollutant-specific (Munõz et al., 2005). Chlorella genus is reported as highly pollutant tolerant microalgae (Palmer, 1969). Heterocystous nitrogen fixing blue green algae can be used for treatment of nitrate waste and production of nitrogen fertilizer (Benemann, 1979). These are filamentous algae consisting of two types of cells: the heterocysts, responsible for ammonia synthesis and vegetative cells, which exhibit normal photosynthesis and reproductive growth. Benemann (1979) isolated the sewage effluents adapted algae and cultivated them in small ponds. Significant rates of biomass production and nitrogen fixation were achieved. Jaag (1972) reported that in Switzerland, in the waste water treatment plants first organic pollutants are decomposed i.e. mineralized and phosphates and nitrates etc. are voluntarily discharged into rivers and lakes; which thereby become over fertilized. This "eutrophication" is manifested by a dense production of filamentous and planktonic algae. At the end of vegetation period, this mass of organic matter dies and causes sedimentation at the bottom of lakes as partly undigested sludge. Eutrophying substances may be eliminated by chemical precipitation (phosphorous) and biological oxygen reduction (nitrogen) in sewage treatment plants.

Table 1.4 Algal-bacterial/microalgal consortia for organic pollutant removal

Compound	Reactor	Conontuim	Removal rate (mg/l/day)	Reference
Acetonitrile	600 ml Stirred Tank Reactor (STR)	*C. sorokiniana/bacterial consortium*	2300	Munoz et al., 2005a
Acetonitrile	50 l column photobioreactor	*C. sorokiniana/bacterial consortium*	432	Munoz et al., 2005b
Black oil	5 ml tubes	*Chorella/Scenedesmus/ alcanotrophic bacteria*	—	Safonova et al., 1999
Black oil	100 l tank	*Chorella/Scenedesmus/ Rhodococcu/Phormidium*	5.5	Safonova et al., 2004
Phenanthrene	2-l STR with silicone oil at 10%	*C. sorokiniana/ Pseudomonas migulae*	192	Munoz et al., 2005c

Contd...

Phenanthrene	50 ml tubes with silicone oil at 20% 600 ml STR with	*C. sorokiniana/ Pseudomonas migulae* *C. vulgaris/Alcalý´genessp.*	576 90	Munoz et al., 2003a Essam et al., 2006
Phenol Phenol	NaHCO3 at 8 g/l 100 ml E-flasks	*Anabaena variabilis*	4.4	Hirooka et al., 2003
Salicylate	600 ml STR	*C. sorokiniana/ Ralstonia basilensis*	2088	Munoz et al., 2004
p-Nitrophenol	—.	*C. vulgaris/C. pyrenoidosa*	50	Lima et al., 2003

Biofuel

Yet another important use of microalgae is biofuel production. Concept of using microalgae as a source of fuel is much primitive. They are remarkable and efficient biological factories capable of utilizing a waste form of carbon (CO_2) and converting it into a high density liquid form of energy (natural oil). This ability has been the foundation of research program of biofuel production from microalgae. Initially efforts were directed towards the direct combustion of algal biomass for production of heat and steam. Presently, research is focused on microalgae, which are particularly rich in oils for diesel production and whose yield is considerably higher than that of conventional sources like sunflower or rapeseed. Microalgae offer several advantages over terrestrial plants. Their photosynthetic efficiency (6–8%) is much higher than that of terrestrial plants (1.8–2.2%). Moreover, they can easily adapt to a wide range of pH and can grow in fresh or marine water. There are three main options of fuel production, which include, methane gas via thermal or biological gasification, ethanol via fermentation and biodiesel. Another major attraction is their exceptional capacity of assimilating CO_2. This has another potential application. If algal ponds are constructed next to electric or coal based power stations, the CO_2 emissions can be utilised by the algae. This biomass can be used for biofuel generation (Brown and Zeiler, 1993). Finally, molecular biology aspects can also be applied to engineer the algae for enhancement in the area of biofuel production.

Chapter 2

CULTIVATION

2.1 HISTORY OF MICROALGAL MASS CULTURE

At present, micro algae are cultivated for food or for metabolites. The history of microalgal cultivation dates back to the Aztecs (Sommer, 1988), where people mastered cultivation of these algae and incorporated them as food. Microalgae are rich in protein, carbohydrates, fatty acids, amino acids, trace elements, and vitamins (Waaland, 1981). The high protein content of these micro algae implies that these could be potential source of feature food and feed. The advantages of culturing microalgae as a source of biomass are many (Vonshak, 1990).

- Microalgae are considered to be a very effective biological system for harvesting solar energy for the production of organic compounds via the photosynthetic process.
- These are nonvascular plants lacking complex reproductive organs, making the entire biomass available for harvest and use.
- Many species of microalgae can be induced to produce particularly high concentrations of chosen, commercially valuable compounds such as proteins carbohydrates, lipids and pigments.
- Microalgae undergo a simple cell division that enables them to complete their lifecycle faster compared to higher plants. This allows a more rapid development and demonstration of production processes than with agricultural crops.
- For many regions suffering low productivity due to poor soils or shortage of sweet water, farming of microalgae that can be grown using sea or brackish water may be almost the only way to increase productivity securing a basic protein supply.

Microalgal biomass systems can be adapted to various levels of operational or technological skills, from simple, labor-intensive production units to fully automated systems requiring high investments. First commercial production systems for microalgae were developed in Japan (Tsukada et al., 1977). In 1960's many chlorella cultivation facilities (circular ponds) were established in Japan to produce dried algal powder. During 1980's there are as many as 40 production units producing 1000 kg of chlorella biomass per month (Kawaguchi, 1980). In 1995 a large (300 ton/yr) plant was started in Indonesia, increasing world production almost by 50%. As compared to *Spirulina, Chlorella* cultivation is difficult mainly because of contamination problems and the need for expensive centrifugation for harvesting. The next commercial development was production of blue green alga *Spirulina. The* alga is considered as a traditional food in Mexico and Africa. Large-scale production was started in 1970's near Mexico City, in a carbonate rich pond and during 1980's in the U.S., at plants in California and Hawaii. *Spirulina* is used basically as food supplement, with applications in aquaculture, feed and as a food colorant. At present the dominating producers in the world are the production facilities in California and Hawaii, with about 350-tons/year each. In India, the first commercial microalgal (Spirulina) cultivation started in 1986 at Severiarpuram, Tamilnadu. The initial production was around 10 tons/annum; currently it is under the management of New Ambadi of Murugappa group. Another installation for phycocyanin production was established at Wardha, by Cyanotech Byproduct Pvt. Ltd. The capacity of this unit is around 6.8 ton/year (Venkataraman, 1978). The most recent commercialized microalga is *Dunaliella Salina.* Few strains of *Dunaliella salina,* grow at high salinities, producing high quantities of beta-carotene. It is grown in high salinity ponds in California, Hawaii, and Australia. The carotenoid is used widely as food colorant and vitamin/antioxidant food supplement.

2.2 PRODUCTION SYSTEMS

Microalgae are grown commercially in different ways: (1) open cultivation using natural sunlight, (2) closed cultivation using natural light and (3) closed cultivation using artificial illumination. Each system has its own advantages and disadvantages. The selection of the production system depends on the degree of controlling parameter required to produce the desired metabolite and the value of the product. Fig. 2.1(a) and (b) shows a typical open and closed microalgal cultivation system.

Fig. 2.1 (a) Open ponds and mass recovery system
(Source: http://www.spirulinasource.com/

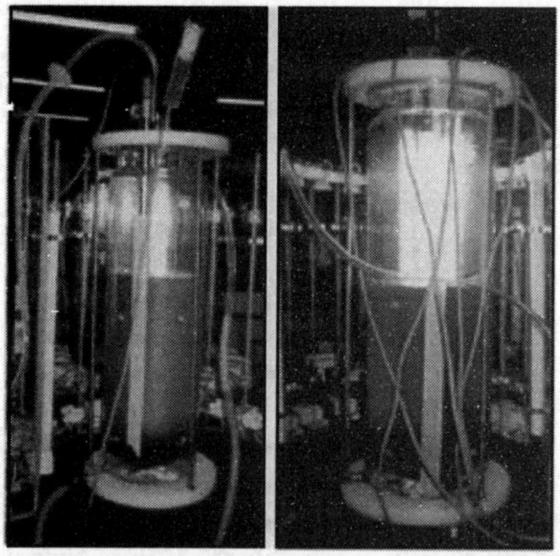

Fig. 2.1. (b) Closed type of algal cultivation
(Source: Photographs from authors research lab, UICT)

Open ponds

Many designs of open systems utilizing natural sunlight have been proposed and constructed (Chaumont 1993; Phushparaj et al., 1997). These cultivation systems are large open ponds or raceways. The major advantage of these designs is that the light energy is free. However, these cultivation systems have significant disadvantages. In open systems contamination by other organisms is the major problem often associated with these systems. The problem of contamination has been solved to some extent by culturing algae that require special growth conditions that would prevent the contaminating microorganisms; however, by doing so (i.e. facilitating species specific conditions) limit the usefulness of these systems to a small number of organisms. Microalgae, which tolerate unique conditions, include *Dunaliella* (grown at very high salinity) and *Spirulina,* which can be grown at high pH. Low productivities are often seen associated with open culture systems, which need to process large quantities of water to harvest the algae. Also out door open systems are subjected to daily and seasonal changes in light intensity and temperature, which makes it very difficult to control operating parameters and reproduce specific culture conditions. However, for some micro algal metabolites this technology has proved very successful, producing tons of dried biomass per year (Lee, 1997). This is especially true for Spirulina, which is largely cultured in the U.S, Mexico, Thailand and China (Metting, 1996, Li&Qi, 1997 and Vonshak 1997). Studies on open systems have been reported for *Spirulina platensis* (Richmond et al., 1990) and *Chlorella* (Doucha & Livansky, 1995). The current variants in open system designs are shallow ponds, tanks, circular ponds and raceway ponds. All the designs cited have their own advantages and disadvantages. The selection of a particular system however depends on the algal characteristics, climatic conditions, availability of water and cost of the land. In cultivation of *Dunaliella. salina,* Betatene Ltd., employs ponds having nearly 250 ha of area at their production facilities at Hutt Lagoon and Whyalla. These systems however have no external mixing facilities and relay on wind and convection. Employing such vast amount of land for cultivation is feasible because of low cost of land, free and clean water (seawater) and supporting climate all around the year. In addition the company has a good harvesting system (Borowitzka, 1992). However, when the cultivation systems in USA and Israel are considered because of higher land cost and additional requirement of sodium chloride (15–25% w/v) for maintaining high salinity for Dunaliella, they employ paddle wheel based mixing in raceway ponds for accomplishing higher cell densities (Ben-Amotz, 1995). These precautions (low pond area and paddle wheel based mixing) maximize cell densities and make the cultivation process feasible economically in these countries. Similar raceway systems with paddle wheel based mixing are used for *Spirulina* and *Chlorella.* The blue green alga (*Spirulina*) however is resistant to contamination particularly when grown in bicarbonate enriched medium; *Chlorella* is more prone to contamination because of its cultivation pH being close to neutrality. Chlorella cultivation require clean atmosphere in order to prevent contamination with other unwanted organisms. The Caracol at Sosa Texcoco in Mexico is the only successful largest cultivation system and is basically a spiral shaped solar evaporator of 3200 m diameter with a surface area of 900 ha. This system however was abandoned because of increased contamination as the result of industries in the near by places. The increased risk of contamination which is often found in chlorella cultures would mean that they should be grown in batches instead of single large pond with periodic reseeding of pond with new inoculums (Kawaguchi, 1980). In addition, since the climatic conditions do not support the cultivations year around maximum productivities is a must for open systems in order to meet high capital cost for operating these systems. Since a greater amount of light spreads in shallow waters, there exist a relationship between pond depth, culture density and productivity. This unique relation was studied extensively on Spirulina cultures (Vonshak, 1997). These studies are essential for maximizing biomass productivities. In systems like open ponds the algae are limited by CO_2 however, providing pure CO_2 in large scale is not feasible economically. Intermittent addition of CO_2 for Spirulina cultures is essential for maintaining alkalinity (Belay, 1997). Thus the advantageous effect of low cost and simple design of these open systems was over masked by the operation cost, low productivities and possibility of contamination. Keeping in view of more efficient and alternative systems, closed systems are proposed and designed for better productivities and reducing the cost of land and more significantly controlling the environmental conditions when operated indoors.

Closed systems

Closed systems for microalgal cultivation, essentially consist of a typical bioreactor with a suitable arrangement for illumination. Bioreactors may be classified as shown in Fig. 2.2.

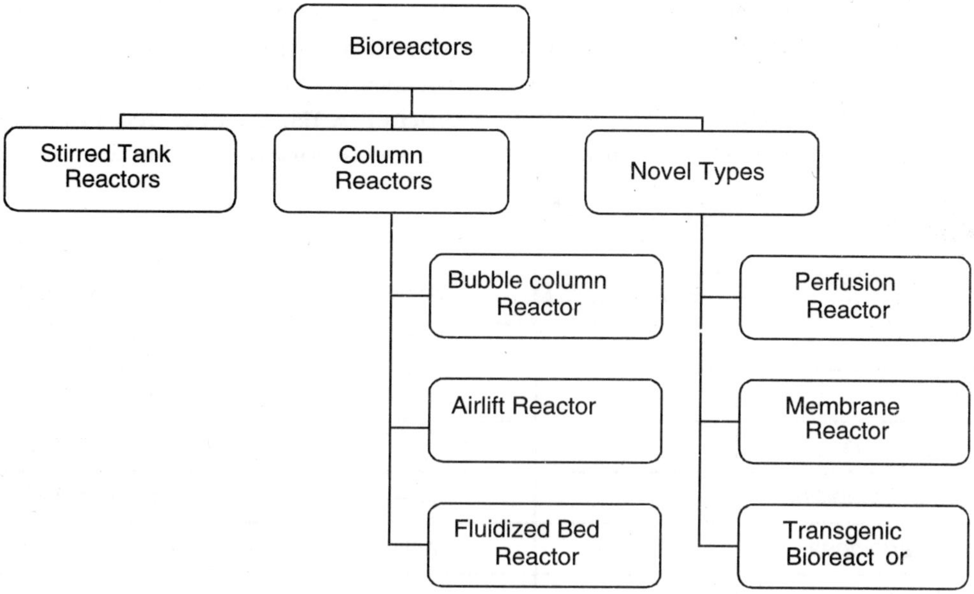

Fig. 2.2. Classification of Bioreactors

Algal cultivations in closed systems placed outdoors have been reported (Richmond et al., 1993, Qiang and Richmond 1994, Molina et al., 1995, Spektorova et al., 1997). In closed systems the alga is enclosed in transparent glass or plastic bags and the vessels are placed outdoor for entrapping sunlight. The closure of the systems prevents the contamination of the culture with other algal species. These systems have higher surface to volume ratios as compared to either ponds or raceways (open systems). This increased surface to volume ratio of the vessels ensures better light entrapment than in open systems. However, the reproducibility of either biomass or metabolite productivities is problematic in out door closed systems as fluctuations in light and temperature is seen with change in season. In addition, outdoor closed systems have problems in removal of oxygen and provision of temperature control. Although these issues can be solved, the cost of such installations can increase the cost advantage of natural sunlight. In an attempt to counteract the problems associated with outdoor systems, several designs have been proposed and constructed for the indoor, closed cultivation of algae using artificial illumination (Ratchford and Fallowfield 1992). These vessels, which are frequently referred to as photobioreactors (PBRs) are in principle akin to conventional fermenters, the major difference are that they are driven by light instead of an organic carbon source. Photobioreactors apart from sharing the advantages (preventing contamination) of outdoor enclosures, provide better controlling atmosphere in terms of light intensity, temperature, aeration, pH and in optimization of other culture parameters. Photobioreactors are generally more expensive to construct than outdoor enclosed systems.

However, the cost can be justified when high value products are targeted from the biomass. Photobioreactors can be used for cultivation of different microalgae (Ratchford and Fallowfield, 1992); they can be vital tool for small-scale production of high value metabolites, such as stable – isotope-labelled biochemicals, unsaturated fatty acids etc. Closed systems investigated for some microalgae are reported in Table 2.1.

Table 2.1 **Closed algal cultivation systems**

Alga	Bioreactor	Orientation	Reference
Spirulina maxima	Tubular	Horizontal	Torzillo *et al.*, 1986
Spirulina sp	–	–	Torzillo *et al.*, 1986
Spirulina platensis	–	–	Richmond *et al.*, 1993
Porphyridium cruentum	–	–	Chaumont, 1993
Chlorella pyrenoidosa	–	Inclined	Lee *et al.*, 1995
Chlorella pyrenoidosa	–	–	Lee & Low, 1991
Tetraselmis chuii	Coil	Vertical	Borowitzka, 1992
Isochrysis galbana	–	–	Hu & Richmond, 1994
Spirulina platensis	Flat plate	Inclined	Hu & Richmond, 1996
Spirulina platensis	–	–	Hu & Richmond, 1996
Spirulina platensis	–	–	Tredici *et al.*, 1997.

Based on the above discussion it can be seen that both the open and closed mode of cultivation have a number of advantages and disadvantages. Hence, the choice should be based on some important parameters. These are presented in Table 2.2.

Table 2.2 **Open Vs Closed Systems**

Parameter	Open ponds	Closed systems
Contamination risk	Extremely high	Low
Space required	High	Low
Water losses	Extremely high	Almost none
CO_2 losses	High	Almost none
Biomass quality	Not susceptible	Susceptible
Variability of cultivation	Restricted to few algal sp.	Variation is possible

Contd...

Flexibility of production	Difficult	Very flexible
Reproducibility of Parameters	Dependent on exterior Conditions	Possible within certain Tolerances
Process control	Not possible	Possible
Ease of Scale-up	Not possible	Possible
Weather dependence	Absolutely dependent	Insignificant
Cultivation time	Long, approx. 6–8 weeks	Relatively short, 2–4 weeks
Biomass concentration	Low, approx. 0.1–0.2 g l^{-1}	High, approx. 2–8 g l^{-1}
Efficiency	Low	High

2.3 PARAMETERS INFLUENCING CULTIVATION

The basic ability of photoautotrophic microorganisms to utilize light quanta in the visual range as an energy source for metabolism is handicapped by high population densities. As against heterotrophic microorganisms, such as yeasts or bacteria where the distribution of the organic molecules as energy carriers by mixing is solved technologically, the supply of photons is dependent on surface area. Light dissipation in the cell suspension is realized by numerous quantum interactions with exponential decrease along the boundary normal. Only the conversion process with the assimilation pigments of the microorganisms are used as energy receptors for the development of biomass, i.e. the radiation energy is lost largely as heat, fluorescence etc. The large energy density at the boundary surface contrasts with the insufficient supply to the cells at only a few mm layer thickness. Despite a few commercial applications a significant break through has not occurred. This is mainly due to the large space requirements for open plants and the expensive cultivation technique for closed plants, which allows only for a relatively low biomass concentration of 1-2 g dry matter per liter down to 0.4 g in open systems. Mainly extremophilic species of low contamination risk like osmotolerant *Dunaliella* are cultivated. Nearly all the commercial plants are based on open pond technologies. These systems, however, seem to have reached their limits. The gap between the theoretical biological potential of microalgal biomass and the productivities actually achieved may be reduced by developing closed cultivation systems (Richmond et al., 1993). It is rather difficult to compare the open pond technology with closed systems and indoor photobioreactors because of the strong variations of the prevailing boundary conditions. The general review suggests that open systems are predominating in mass cultivation because of cost considerations, and photobioreactors become more and more interesting for the preparation of valuable substances and for special applications (Weisman et al., 1988).

Influence of light

Microalgae have an extremely high productivity compared to higher plants. Some strains of high-temperature adapted green algae, e.g. *Chlorella pyrenoidosa* are able to double their cell number every 2½ hours. As opposed to higher plants, photosynthetically inactive cells (such as roots or trunks) are missing and microalgae are a favored subject of research for the determination of complex photosynthetic processes, especially with regard to quantum yield.

Photosynthetic production of biomass is generally represented by the following equation

$$\alpha\, CO_2 + \beta\, H_2O + \gamma\, NH_3 + \text{minerals} + \text{n hí} \longrightarrow \alpha\, [\text{CHNO minerals}] + O_2$$

Where n hí is the number of energy quanta of a certain frequency, which is required for the production of 1 mol O_2. Effects such as dark respiration or the energy required for metabolism maintenance will reduce the yield of fixed carbon with regard to the oxygen quantity produced. In the classical model of photosynthesis, the Z-scheme, 8 mol quanta absorbed are required for the evolution of 1 mol O_2, at which approximately 3 mol ATP are produced, cyclic photophosphorylation included. They are also required for reducing 1 mol CO_2 to glucose in the dark reaction. The real requirement of quanta for photosynthesis, however, is controversial and given by different authors as 6–16 quanta. The data of the calculated photosynthetic efficiency (PE), the quotient of the energy accumulated in the biomass and the light energy absorbed, are correspondingly different. A theoretical maximum PE of approx. 29% is calculated at 8 mol light quanta (ë = 575 nm) per 1 mol fixed CO_2. With regard to the solar spectrum the conversion of the irradiated sunlight is corrected theoretically to 18%. This value, however, remains questionable under real cultivation conditions, all the more since the real conversion rates do not exceed 10%. After extensive examination, Oswald (1988) calculated a maximum utilization of sunlight of approximately 5-6% for outdoor plants, which was also reached in a continuously mixed culture in Richmond, California (Oswald, 1988). Algae, therefore, by far exceed agricultural crops, such as maize with a PE of 1.3%, fast growing tropical plants as sugar cane or bullrush millet with up to 4.3% and are even superior to technical equipment such as classical photocells or solar collectors. Furthermore, a number of light-regulated mechanisms are available, which make the metabolic cycles of photosynthesis more flexible, although, requiring additional energy. The information and energy content of light influence the thioredoxin system or the development of secondary messengers. Thus, quantity and quality of the light are of decisive influence to a defined cultivation of phototropic microorganisms.

Nutrition

The type of medium employed in microalgal cultivations plays an important role in productivities. The medium composition influences the biochemical composition of the cell. Apart from carbon and nitrogen requirements, the alga requires other minerals and trace elements in sufficient quantities for balanced growth of the cell.

Carbon requirement

Depending on carbon and energy source the algae are classified as below:

- **Photoautotroph**

Photoautotroph requires an inorganic carbon (CO_2, $NaHCO_3$, and Na_2CO_3) as a carbon source and light as energy source. Most of the blue-green algae appear to be obligate phototrophs and are unable to grow heterotrophically in the dark. The micro algae have concentrating mechanisms for CO_2 as well as bicarbonate by which the algae are able to survive and grow autotrophically with the least available concentration of inorganic carbon.

- **Heterotroph**

Organisms which utilize organic carbon (glucose, acetate etc.) as sole carbon and energy source are heterotrophs. Unlike phototrophs the requirement of light is excluded for these organisms. Although most of the blue green algae are obligate photoautotroph, some algae show both phototrophy and heterotrophy. Algal strain showing aerobic heterotrophic growth are: *C. sorokiniana, C. regularis, C. pyrenoidosa, Scenedesmus, Haematococus, Spirulina, Nitzschia laevis, Chlamydomonas reinhardtii, Scenedesmus obliquus, Synechocysti and Plectonema.*

- **Mixotroph**

Mixotrophic organism utilizes organic carbon as carbon source and inorganic elements as electron donors. According to some authors (Marquez et al., 1993 and Hata et al., 2000) in mixo-trophic culture, there is possibility of simultaneous uptake of organic carbon and CO_2, as carbon sources for cell synthesis, there after the CO_2 released via respiration is trapped by the alga and reused in the presence of light. Thus mixotrophs acquire energy by catabolizing organic compounds via respiration and converting light energy into chemical energy via photosynthesis. Mixo-trophic culture yields in terms of biomass or pigment production is known to be several times higher than the photo autotrophic cultures (Marquez et al., 1993). Not all the algal members are capable of metabolizing organic compounds for mixo-trophic growth nevertheless microalgae such as Spirulina (Marquez et al., 1993),Chlorella (Endo et al., 1977), Scenedesmus (Shamala et al., 1982), and Haematocccus (Kobayashi et al., 1992) are capable of utilizing organic compounds as carbon sources.

Nitrogen sources

Chief nitrogen sources utilized for algal growth are ammonia, nitrate and nitrogen gas (Fogg, 1973). Essentially all the microalgae metabolize nitrate and ammonia, moreover algae and bacteria incorporate ammonia very rapidly. Further various dissolved organic nitrogen can be metabolized by the algae also help in removing ammonium ions adsorbed to suspended particulate matter. The elemental nitrogen is metabolized by only certain group of blue green algae and they can do so only under limited environmental conditions (Fogg, 1973; Bold and Wynne, 1978). Most of the inorganic nitrogen sources are well suited for microalgal cultivation however the proper selection of nitrogen source depends on the type of metabolite of interest. Algal cultivations aiming at producing high biomass, chlorophyll, urea and ammonium compounds are regarded as the best nitrogen sources. However, due to their toxicity and volatility of these compounds at higher temperatures, these compounds are less preferred in batch cultivations. Addition of pulses of these compounds using different protocols in fed batch mode can reduce the evaporation loss of these compounds and enhance growth of the alga (Danesi *et al.*, 2002). Similarly, for production of pigments like phycocyanin from blue green alga, either sodium or potassium nitrate sources are well suited. For fatty acid production though urea is reported to give high content (Mahajan and Kamat, 1995), the reproducibility of yields is doubtful. Thus the choice of nitrogen plays important role and it primarily depends on metabolite of interest.

Minerals

Microalgal growth is often limited by availability of certain macronutrients like potassium, magnesium and calcium in the medium. Apart from nitrogen, microalgae require phosphate in large quantity. It is required for the maintenance of osmotic pressure and cell size, thereby influencing photosynthesis and energy production. Magnesium is an important metal; it forms an integral part of chlorophyll and helps directly in photosynthesis. Although, calcium is not essential for algal motility, its addition to the chemo-taxis medium stimulates positive chemotaxis. Calcium is well known to regulate chemotactic behavior in bacteria, and it was proved in *D. tertiolecta* that the algal chemotaxis towards ammonium ion increases in the presence of calcium (Roy, et al., 1978).

Trace elements

The trace metals, essential for micro algal growth are integrated into coenzyme factors, which participate in photosynthetic reactions. Among the metals, the concentrations of Fe, Mn, Zn, Cu and Co and sometimes Mo and Se are seen limiting the algal growth in natural waters. Molecules which form complex with metals (chelating agents) are thought to influence the availability of these elements. Chelators can be regarded as metal buffers, in controlling concentrations of free ionic metal. The free ionic metal plays a vital role in either promoting or inhibiting algal growth. Proper chelation is essential to maintain the free trace metals in concentrations below toxic level to the microalgae. Metallic ions such as iron tend to precipitate and become unavailable to microalgae. In natural seawater, dissolved organic molecules (generally present at concentrations of 1–10 mg l^{-1}) act as chelators. A well known chelator most widely used in algal culture medium is EDTA (ethylene-diamine-tetra-acetic acid), keeping in view the larger amount of Ca and Mg in sea water, algal medium preparations should have larger amount of this chelator in order to form complexes with these metals. Apart from chelating behaviour, EDTA reduces precipitation particularly during autoclaving. However at high concentrations EDTA has been also reported to be toxic to microalgae. Organic chelator (citrate) can also be used. The ratio of chelator: metal in culture medium ranges from 1 : 1 to a maximum of 10 : 1 in various media. High ratios however result in metal deficiencies due to excess chelation and hence an intermediate ratio is recommended.

Environmental factors

Apart from nutritional requirements of microalgae, several environmental factors affect its cultivation. These are discussed below:

pH

Microalgae, particularly blue-green algae found in natural alkaline waters and those that are grown at laboratories under high alkaline conditions in bicarbonate enriched medium exhibit high growth rates. Since bicarbonate is the predominating carbon ionic species among dissolved total inorganic ions at an alkaline pH range ($7 > pH < 10$), the ability of these group of algae to grow in alkaline atmosphere indicates that they can metabolize bicarbonate more efficiently for photoautotrophic carbon fixation. The control of pH in microalgal culture media assumes significance since certain algae can grow only within narrow pH ranges. In addition, it helps in preventing the formation of precipitates. Most often the culture medium is prepared using natural seawater. The pH of seawater is around 8 and is maintained constant because of the buffering capacity of natural seawater (due to a bicarbonate buffering system). Alternatively this buffering system can be achieved in medium prepared with ordinary tap water with bicarbonate (Zarrouk medium, SOT medium). The buffer system is changed during autoclaving, as high temperature drives out CO_2 in the medium which finally results in increasing the pH due to shift in bicarbonate buffer system. This behavior is akin to the high density culture of microalgae where larger amount of CO_2 is taken out from the medium during photosynthesis which results in pH rise of the medium at the end of fermentation time especially in batch cultures. However, when the medium cools after autoclaving, CO_2 in the atmosphere enters the medium controlling the pH to some extent. Loss in pH can be compensated if some of the following measures are taken. The pH of seawater/bicarbonate enriched tap water can be lowered prior to autoclaving (adjusting to pH 7–7.5 with 1 M HCl) so that any increase in pH is compensated. Additions of small amounts of buffers like bicarbonate, Tris (Tris-hydroxy-methyl-amino-methane), or glycylglycine can be used as supplemented buffer to maintain pH in naturally buffered medium. These additions are generally not necessary if media are filter sterilized. The CO_2 loss and subsequent drop in pH can be prevented in dense cultures by having a large surface area of media exposed to the atmosphere when compared to the volume of the culture, or by bubbling with either air or air enriched with CO_2 (0.5–5%). Higher concentrations of CO_2 in the absence of large biomass causes a significant decrease in pH. In such a case excessive aeration is not an option for majority of microalgae as physical disturbance may inhibit growth or damage shear sensitive cells.

Temperature

Majority of algae grow at higher temperatures generally between 30–37° C. Among blue green algae, the optimum temperature for *Spirulina platensis* is between 35–37° C. Similarly, the Chiorphyte (*Haematococcus*) grows best at 20° C. Few species of *Dunaliella* are known to have temperature tolerance up to 40° C. Apart from functional role of growth, temperature is a stress parameter for certain algae like Spirulina where this physical parameter can be used to enhance essential unsaturated fatty acids. The alga accumulates larger quantity of unsaturated fatty acid (GLA) at lower temperature to increase membrane stability.

Light

In contrast to common chemical nutrients it is not possible to distribute light homogeneously. Illumination is the most important factor influencing biomass composition, growth rate, and product formation. The alga uses light which is well within photosynthetically active radiation (i.e. 400 – 700 nm).The penetration of light inside dense algal cultures is a major problem which needs greater attention in microalgal cultivations. In addition, in high density cultures, mutual shading is a factor which reduces light availability to other algal cells. These problems have to be taken into consideration in the design of photobioreactors for micro algal cultivation. In general it is always desirable to have larger surface to volume ratios of the cultivation systems (photobioreactors) for entrapping maximum light energy.

2.4 PHOTOBIOREACTOR DESIGNS

A lot of research is presently being focused on closed cultivation systems for microalgae in photobioreactors. Several such reactors have been designed and studied. Some of these are discussed below.

Bubble column

Vertical columns such as bubble column and airlifts are compact reactors. They are cheap and can operate mono aseptically (Chisti, Y., 1989). These reactors are preferable option for production of biomass in large amount. Hardly few studies have been done in characterization of microalgal cultures in bubble columns (Fig. 2.3 (a), (b). Bubble columns have higher mass transfer coefficients and relatively good hydrodynamic properties among air sparged reactors. The hydrodynamic behavior of bubble columns deviates largely in terms of liquid circulation with that of airlift reactors. The liquid circulation in airlift depends on the gas flow rate while in bubble column the liquid flow is independent of the gas flow. Adequate knowledge exists on bubble columns regarding non-phototrophic cultivation applications (Chisti and Moo-Young, 1987; Chisti and Moo-Young, 1989; Chisti, 1998, 1999 a, b; Deckwer 1992; Joshi et al., 1990; Silva et al., 1987; Suzuki et al., 1995). Sanchez et al., (2000) characterized bubble column in terms of hydrodynamic properties and mass transfer capabilities for phototrophic applications. The bubble column reported (Sanchez. et al., 2000) is a transparent poly methyl methyacrylate with lower 0.25 m sections made of stainless steel. The vessel has a 0.193 m diameter and 2.3 m height with a working volume of 0.06 m^3. The sparger is a cross shaped pipe sparger, the width of each blade is 1.6 m. The blade has a total of 17 holes each of 1 mm diameter. The performance of the reactor was evaluated by growing the culture of Phaeodactylum tircornutum. In the batches tested the algal biomass concentration reached 4 kgm^{-3} after 260 h. The maximum specific growth rate was 0.022 h^{-1} at a specific power input of 109 w m^{-3}.

Fig. 2.3. (a) *Bubble column type photobioreactor*
Halide light illumination from outside the reactor, inside the enclosure
*(**Source:** Photographs from authors research lab, UICT)*

Fig. 2.3. (b) *Bubble column type photobioreactor (inside view)*
Halide light illumination inside the enclosure
*(**Source:** Photographs from authors research lab, UICT)*

Airlift reactor

An airlift reactor (ALR) is a pneumatically agitated device consisting of a pool of liquid which, is divided into two distinct zones, a riser and a down comer (Fig. 2.4 a, b). Differences in gas holdups, in riser and down comer results in bulk density difference of the liquid which causes circulation of fluid in the reactor (Chisti, 1989). In airlift, the volumes enclosed by annulus (riser) and draft tube (down comer) can be regarded as illuminated and dark regions, respectively. This defined circulation pattern gives light/ dark cycles for micro algal cell cultures influencing the productivities in the reactor (Phillips et al., 1954)The liquid circulation velocity determines the frequency of light and dark cycles. The liquid circulation velocity in turn depends on design of reactor and superficial gas velocities. Higher linear liquid velocities are accomplished in airlifts with out external recirculation mechanism, moreover the operation range of gas and liquid velocities in airlifts are broader than the bubble columns (Weiland and Onken, 1981). Mixing in the ALR can be accomplished without causing too much shear force in the liquid phase, which can inhibit the algal growth. Geometrical variants exist in airlift reactors. (Chisti, 1989) These include external loops, split cylinders, and rectangular airlifts. Earlier studies by Sanchez et al., (2000) showed that airlifts can be potential enclosed systems for algal cultivations, since then many researchers studied the feasibility of algal cultivations in airlifts reactors (Sanchez et al., 2002; Xiaoxi, 2004; Wu and Goosen., 1995; Sontaya et al., 2005).

Fig. 2.4.(a) *Airlift photobioreactor with external circulation External illumination by fluorescent tube lights*
*(**Source:** Photographs from authors research lab, UICT)*

Fig. 2.4.(b) *Airlift photobioreactor with internal circulation External illumination by fluorescent tube lights*
*(**Source**: Photographs from authors research lab, UICT)*

Tubular systems

Tubular photobioreactors consist of long transparent tubes with diameters ranging from 3–6 cm and lengths ranging from 10–100 m. In horizontal tubular photobioreactors apart from the mixing assisted by circulation of fluid in the tubes, an external air bubbling using airlift are best accepted (Gudin and Chaumont., 1983). The horizontal tubular systems besides not having considerable advantages compared to conventional open ponds have severe limitations in terms of temperature control, need for frequent re-carbonation, build up of inhibitory oxygen levels, foaming, fouling etc. Additional costs are incurred in order to rectify these limitations and these costs are justified only when the targeted products are of high value. (Weissman et al., 1988). Keeping these limitations in mind an airlift used in conjunction with the tubular system would be a better option which gives good mixing, better gas exchange and removes dissolved oxygen. A significant variation among tubular systems, is the design of the coiled type. The design was first used by Davis et al., (1953) and later it was adopted by many researchers (Watanabe et al., 1995; Tredici et al., 1997). The design in general includes PVC tubes (3 cm) wound on a rigid vertical support. The height of the coil varies and that used by Tredici et al., (1997) is 1.65 m. The free end of the tubes at the top was connected to a degasser which in turn connects with a single down flow pipe with the bottom of each tube. This makes a continuous recirculation flowing system. Air is bubbled at the bottom of the reactor where the bubbles rise along the length and escapes at the top of the degasser. The coiled systems facilitates the operation of long tubes placed on relatively small area. Carlozzi et al., (1996) studied and compared strongly curved tubular design (CTP) with normal straight tubular design (STP). The STP is similar to the one described by Torzillo et al., (1993). The cultures were recycled with the aid of peristaltic pumps. The productivities of Spirulina platensis culture was studied in both reactors with respect to flow rates. With the increase in flow rates the productivities of Spirulina platensis increased significantly in CTP as compared to STP. According to the authors, better performance of the CTP over STP was due to the intermittent illumination as a result of secondary flow motion generated in the bends. In addition, they noticed a significant change in power consumption for inducing similar flow rate. A typical tubular photobioreactor is shown in Fig. 2.5.

Fig. 2.5. *Tubular Photobioreactor Illumination by solar rays*

(Source: *Photographs from authors research lab, UICT)*

Flat plate reactors

Flat plate reactors in general consist of a rectangular transparent box with depths ranging from 1 to 5 cm. The height and width of the design varies and in practice these dimensions are less than 1 m. Mixing in flat plate reactor is accomplished by air sparging. The superficial gas velocities range from 0.030 to 0.035 m s^{-1}. The biomass concentrations of Spirulina cultures in flat panel reactors are supposed to be in the range of 5 to 17g/l (Hu et al., 1998; Hu and Richmond, 1994). A variation seen among flat plate reactors is the flat and curved chamber reactors (Tredici et al., 1997). The flat chamber reactor as described by the authors was a rectangular chamber made of a Plexiglas sheet; the chamber was open at the top end. There were no internal partitions provided. The curved chamber reactor, though built on a similar plan, the sheet pieces used to build the front and rear walls had been molded into a nearly hemispherical dome. Anatoly et al., (1996) studied Spirulina platensis cultures in a flat panel photo bioreactor. Mixing was facilitated by 2% CO_2 enriched air. Cooling conditions were accomplished with sprinklers along the top of the front panel.

Gas induced mechanically agitated photobioreactor (GIMAP)

This design is least explored for algal cultivations among gas liquid agitated contactors. Kshirsagar et al., (2002) have demonstrated good mass transfer capabilities of GIMAP. The GIMAP design essentially consists of a cylindrical baffled borosil vessel fitted with a hollow shaft impeller (Fig. 2.6). A gas port at the top of the hollow impeller, facilities simultaneous gas induction and mixing in the reactor. The gas induction in the liquid occurs at a certain critical speed of the impeller. Mixing is accomplished with an over head stirrer. The power consumption for operating these reactors is relatively low as compared to conventional two phase stirred tanks. The usage of gas inducing impellers (Joshi and Sharma, 1977) eliminates the need for separate air sparging and this results in lower, power consumption. In an interesting study using GIMAP, Kshirsagar *et al.*, demonstrated that phycobiliproteins from spirulina plantentis could be extracted without cell lysis. This was attributed to the controlled shear that exists during the algal growth due to which the release of phycobiliproteins was facilitated with out complete lysis of the cell. The isolation of phycobiliproteins and subsequent recovery of cell biomass showed 25% of intact proteins. The direct isolation of phycobiliproteins in GIMAP along with recovery of exhaust cell biomass eliminates the need for a separate downstream process for pigments and thus results in cost effective isolation of phycobiliproteins.

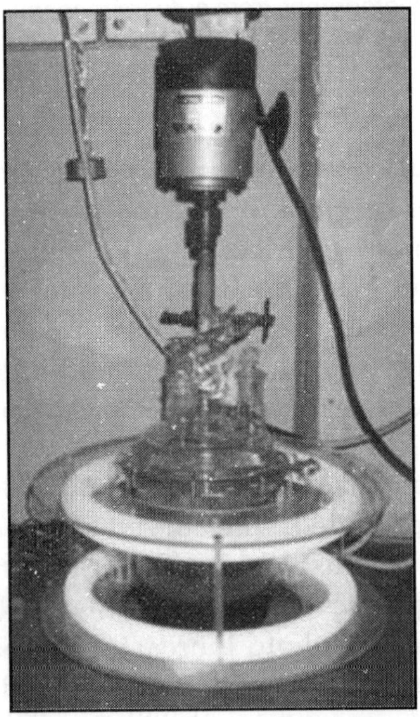

Fig. 2.6. *Gas induced mechanically agitated photobioreactor (GIMAP) Illumination by circular fluorscent tube light (**Source:** Photographs from authors research lab, UICT)*

Novel Designs

Sato et al., (2006) reported some interesting novel designs of photobioreactor; the first of its kind is a dome shaped reactor. In their studies the authors used a computational mixing dynamic study (CFD) to evaluate mixing performance of the reactors. Both mixing and provision of CO_2 is facilitated by aeration. Air bubbles were generated by six nozzles located at an equal interval (60°). CO_2 enriched air is supplied through the six nozzles. The unique feature of the dome is that the reactor is provided with a small device called 'train'. This particular device moves by repulsion against aeration, circulates at the bottom of the reactor and thus enhances mixing conditions in the reactor. The parabola and the diamond are modification of the conventional panel by removing the corners that have an acute angle. The parabola and the pipe consist of concave walls. It has been reported that the pipe geometry showed best performance in terms of productivities for *C. littorale* culture (Sato et al., 2006).

Different models of photobioreactors discussed, exhibit various advantages as well as disadvantages when used for microalgal cultivation. One needs to understand that a photobioreactor suitable for the cultivation of one microalgal species may not give good results when used for another. Some of these *pros* and *cons* are listed in Table 2.3.

Table 2.3 Comparison of typical photobioreactors

Typical photobioreactor design	Advantage	Problems
Tubular Photobioreactor	Plug Flow, High area to Volume ratio	Leakage, Pressure Drop, Biomass sticking to glass
Bubble Column, Stirred Vessel, Gas Induced (Hollow shaft type)	Good Mixing, Low Pressure Drop	Low area to volume ratio, difficult to supply light

2.5 PHOTOBIOREACTOR DESIGN CONSIDERATIONS

Major factors that are considered in design aspects of photobioreactors particularly pneumatically agitated systems include; light source and distribution, gas-liquid mass transfer, mixing and sparger design.

Light source and distribution

For outdoor photobioreactor applications the ultimate source of light energy for phototrophic microalgae is sunlight. However, for an indoor cultivation using photobioreactor technology a wide variety of light sources are reported. The light sources include fluorescent bulbs, xenon lamps and high efficiency light emitting diodes (Lee and Palsson, 1994). Irrespective of the light source, the ultimate goal is to achieve high cell productivities and maximum specific growth rates. Once the light source is decided light availability is an important factor which decides the productivities in photobioreactors. Light intensity decreases exponentially as it penetrates dense algal cultures. Though some parts of the reactor receive light, other zones remain in dark due to optical absorption, larger scattering and self-shading of micro algal cells. However, increasing surface area of the reactor (reducing culture depth) to some extent solves the problem while in the effort of maintaining larger surface area results in lower volumes which ultimately affects the economics of algal cultivations. The light utilization efficiencies can be increased by exposing the algal cells to frequent light and dark periods. Studies on frequent light and dark periods, which is otherwise called flashing light effect is well documented in literature (Phillips and Myers., 1954, Lee and Pirt, 1981; Grobbelaar, 1991). The flashing light effect is mixing induced light and dark cycles.

Gas-Liquid mass transfer

Microalgae requires CO_2 for its growth and at the same time it also requires certain amount of oxygen for carrying out metabolic processes like respiration and production of some metabolites. In order to satisfy the oxygen and CO_2 requirements, the algal cultures are often sparged with air, additionally, the sparging medium (air) is often enriched with CO_2 to increase growth rate of the alga. Thus the rate at which either CO_2 or oxygen is transferred into the reactor depends on mass transfer coefficient of the system. Estimation of Mass transfer coefficient not only gives the performance of the system in terms of oxygen or CO_2 transfer rates but also indirectly gives the estimation of desorption rates of dissolved oxygen in the medium. Relatively high mass transfer coefficients are found in pneumatically agitated bubble column and flat plate reactors than tubular reactors (Hu and Richmond, 1996). During cultivation of high oxygen sensitive phototrophs, the systems with low mass transfer capabilities (horizontal tubular reactors) are less preferred (Babcock *et al.*, 2002). In designing a photobioreactor it is possible to calculate the required mass transfer coefficients to achieve concomitant photosynthetic efficiency at a fixed light intensity. The oxygen production rate (OPR) at steady state is equal to the oxygen transfer rate (OTR). If oxygen production rate is known, the required mass transfer coefficient (k_La) can be calculated by means of the following equation.

$$OPR = OTR = k_La. (C^* - C_L)$$

Where C^* is the oxygen concentration in the liquid phase in equilibrium with gas phase and C_L is the concentration of oxygen in the bulk phase of the liquid.

Mixing

In photobioreactors, good mixing is essential when high cell densities are targeted. Mixing can keep the cells under flotation, removes temperature gradient, helps in improving nutrient mixing and gas exchange. In addition efficient mixing can prevent mutual shading and lower the chances of photo inhibition. Mutual shading can be reduced by mixing in the direction perpendicular to flow, which ultimately increases the number of cells exposed to light. Mixing helps the cells to experience light and dark cycles in the photobioreactor. These short light and dark cycles in fraction of milliseconds are known to increase the productivity of the algal culture (Hu and Richmond, 1996). Appropriate mixing may considerably improve overall light utilization efficiency in two ways: (1) The presence of air bubbles in the medium will increase light penetration depth and (2) Moving of bubbles along the length of the column or reactor will induce certain degree of mixing tangential to the flow direction. Excessive mixing however results in carbon loss in the bicarbonate enriched medium (Rodriguez-Maroto et al., 2005). The medium used in Spirulina cultivations is highly enriched bicarbonate medium which is strongly meta-stable, due to the fact that the bicarbonate medium has CO_2 pressure, which exceeds that of atmosphere. Agitation at increasing rates would mean a permanent loss of carbon and the rate at which it losses CO_2 depends on the intensity of mixing. Agitation commonly noticed in ponds with paddle wheels at relatively low rates may show a slow loss. However intense mixing either with air bubbling or with airlift pumps causes a huge loss of carbon from the bicarbonate medium. Thus continuous air bubbling or aeration in pneumatically agitated systems should be avoided, particularly when the pH of the system is below 10.15. At higher pH (>10.15), the system (partial pressure of CO_2 in bicarbonate solution) will be in equilibrium with CO_2 partial pressure of the atmosphere and this facilitates the CO_2 dissolution in the medium (Daniel de Alava et al., 1997).

Sparger Design

Sparger design is important particularly in pneumatically agitated reactors. The bubble size is influenced by the type of the sparger employed in the systems (Buwa and Ranade, 2002). The sparger types such as sintered, glass, perforated plate, ring, pipe and nozzle sparger, the number and diameter of the orifice and pitch influence the bubble size (Buwa and Ranade, 2002; Polli et al., 2002). The most widely used spargers in bubble columns and flat plate reactors are the perforated and nozzle spargers. In non coalescing systems the sparger design determines bubble size in the reactor (Schafer et al., 2002) However, for coalescing two phase systems (gas-liquid) the influence of the configuration of sparger is limited to the distance from the sparger to the distance in the order of magnitude of the width of the column (Polli et al., 2002). In micro algal cultivations, hydrodynamic stress is a major problem. It was identified that cell death was mainly due to the effect of gas entrance velocities and not as a result of bubble bursting as was thought previously (Barbosa et al., 2003b; 2003). The nozzle diameter and number of holes should be chosen depending on the algal strain so as to operate the reactors below critical gas entrance velocities.

Chapter 3

HIGH VALUE METABOLITES

3.1. HOW AND WHY METABOLITES ARE FORMED?

For over billions of years of evolution, microalgae have uniquely adapted to extreme habitats. Due to their phototropic life they are exposed to high oxygen and radical stresses. This has resulted in the development of numerous effective protective systems against these stress factors. The protective mechanisms are able to prevent the accumulation of free radicals and reactive oxygen species and thus counter cell damaging activities. One such protective mechanism found in microalgae is the accumulation of highly effective antioxidative scavenger complexes, which protect cells from damage by free radicals. For example, *Haematococcus pluvialis, an* unicellular freshwater green algae, when subjected to environmental stresses like nutrient depletion and strong light intensity, forms red cysts as a defense mechanism against the environmental stress (Fig. 3.1).

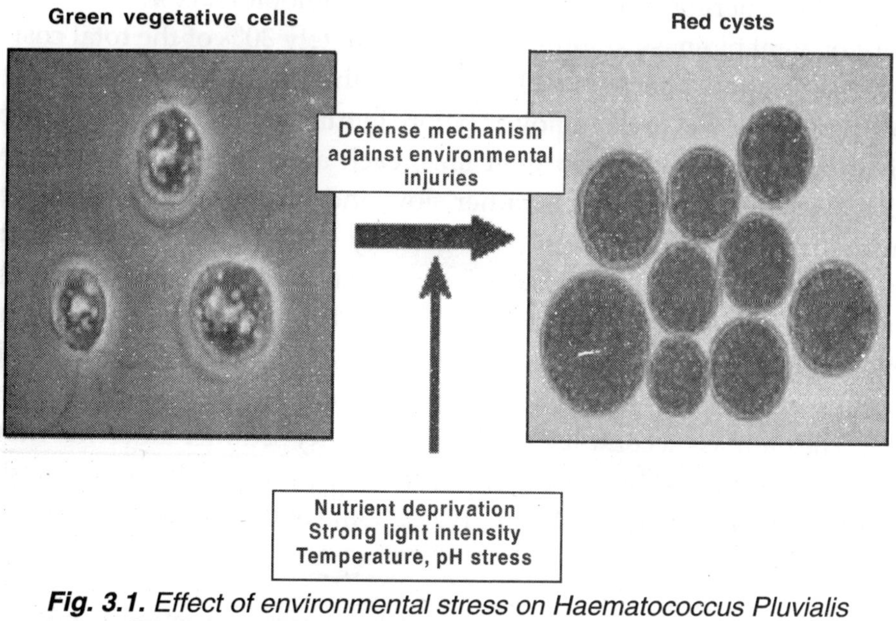

Fig. 3.1. *Effect of environmental stress on Haematococcus Pluvialis*
(**Source:** *http://www.cdrs.re.kr/down/workshop2/21/pdf)*

Yet another example is the accumulation of carbohydrates in microalgae. Besides being a part of many structural components of the cell, these carbohydrates act as reserve food materials (starch and glycogen). Microalgae also accumulate some osmoregulators, which are basically carbohydrates (glucosyl glycerol, sorbitol, mannitol, trehalose etc.).These compounds are accumulated in response to increased osmotic pressure as a result of high salinity in the environment.

3.2. RECOVERY OF BIOMASS

Most algal recovery steps involve filtration, sedimentation and centrifugation techniques. Apart from these steps sometimes flocculation is used effectively to concentrate the cells in the medium. Harvesting of microalgal biomass is a problem because of their small size (3-30mm diameter). Authors studied the cultivation of *Porphyridium cruentum* for production of natural pigments such as phycoetyrhin and phycocyanin in a lab scale stirred tank photobioreactor. It was observed that, cells adhered to the walls of the reactor due to secretion of exopolysaccharides (Fig. 3.2). This led to accumulation of cells at the bottom and sidewall of the reactor. It also resulted in photoinhibition and inhibition of nutrient uptake by the cells due to increased viscosity. Hence there was a decrease in total biomass and phycobiliproteins. Problems like these also add to the difficulties in recovery of biomass. A probable solution could be immobilization of cells on inert support matrices to avoid adhering of cells to the walls of photobioreactor.

Recovery of algal biomass amounts to approximately 30% of the total cost of biomass production. (Gudin and Therpenier, 1986). Since the broths are diluted with medium, larger volume of water has to be removed before getting a concentrate (slurry of biomass) from cultivations ponds. Based on their size, each algae has a different down stream protocol. Recovery using filtration is rather slow and unsuitable at large scale. For very low value products, flocculation and gravity sedimentation are suitable. The sedimentation tanks are akin to those used in waste treatment plants. (Venkataraman, 1978). Centrifugal recovery is the method of choice when high value products are targeted. It is feasible to handle large volumes of broth to recover biomass using centrifuges. The downstream process of the biomass also depends on the final moisture level targeted in the product. The biomass accomplished through gravity sedimentation has more moisture than centrifugally recovered biomass. Greater moisture levels in biomass can show impact on economics of desired product (Mohan, 1978). Some of the common harvesting methods are discussed.

Fig. 3.2. *Porphyridium creuntum biomass found sticking to the reactor (**Source:** Photograph from author's research lab, UICT)*

Flocculation

The fundamental principle behind flocculation is to increase the particle size by concentrating the micro algal cells for easy and effective sedimentation (Elmaleh et al., 1991). Flocculation works on charge difference between flocculent and biomass. Micro algal cells have a negative charge and this prevents them from aggregation or forming flocs. The flocculants carry positive charge, which helps in neutralizing the micro algal negative charge. Many flocculants such as multivalent ions, cationic polymers can be used as flocculating agents. Flocculants should be cheap and effective at low concentrations. Multivalent metal salts are effective flocculants. Most commonly used salts include ferric chloride ($FeCl_3$), aluminum sulfate ($Al_2(SO_4)_3$, alum) and ferric sulfate ($Fe_2(SO_4)_3$). Metal salts like alum is also used to flocculate algae in waste treatment plants (Dodd, 1979). Cationic polymers such as polyelectrolyte serve as alternative to metal slats. Polyvalent organic polymers are effective against Scenedesmus and Chlorella (Golueke and Oswald, 1965). Chitosan, (polymer of acetyl glucosamine), is edible and nontoxic and has been used as an effective flocculant in low concentrations (Richmond, 1986). However its flocculating efficiency is not effective in salt water. Optimal concentration of chitosan for flocculation varies with algal strain (Heasman et al., 2000).

The non toxicity of chitosan has few exceptions where the oyster larvae fed with chitosan showed reduced survival rate of the larvae. In waste water treatment chitosan is used as an entrapment matrix to immobilize microalgal biomass (Kaya and Picard, 1996). In some cases flocculation is achieved simply by changing the pH of the algal medium. In a study on the fatty acid (EPA) recovery from the marine alga skeletoma postpartum, change of pH 10.2 facilitated 80% sedimentation of algal biomass. However the cells were lysed releasing intracellular contents. (Blanchemain and Grizeau, 1999). Auto flocculation by modification of culture medium was also reported (Cannell, 1990). Bioflocculant produced by non-algal origin have been investigated for microalgal flocculation. (Hee-Mock et al., 2001). The bioflocculant, which is obtained from bacterium Paenibacillus sp., proved effective for harvesting Chlorella vulgaris (Hee-Mock et al., 2001).

Centrifugation

Recovery of microalgae from the suspension can be carried out by centrifugation. Centrifugal recovery methods are reported for separation of many biological entities (Belter et al., 1988); however there seems to be a lack of practical guidelines for separation of microalgae using centrifugation. For separation of microalgal biomass using centrifugation the guidelines provided by Chisti and Moo-Young (1991) are relevant. Though centrifugal recovery can be more expensive when treated in large scale, it is most effective in terms of separation of biomass. For applications in aquaculture hatcheries and nurseries in extending shelf-life concentrates (Heasman et al., 2000), centrifugation is considered an ideal method for recovering algal cells (Richmond, 1986). Characteristics of cells like settling behavior, the residence time of the microalgal slurry in the centrifuge and settling depth decide the recovery of biomass from the sedimenting centrifuge. The characteristics of the algal cells in the centrifuge can be controlled by the design and with operating controls, for example the settling depth can be kept small through the design of the centrifuge. On the other hand by selecting appropriate flow rate the residence time of the slurry can be controlled. Many studies have been carried out on the cell recovery and viability of cells at different conditions of centrifugation (Heasman et al., 2000). It has been reported that the viability of the microalgal cells depended on the type of algal species and the centrifugation method (Heasman et al., 2000).

Filtration

The filter presses can be operated under pressure or vacuum. Though they are effective in recovering relatively large microalgae such as *Spirulina platensis* they are not applicable for small size organisms of bacterial dimensions (e.g., *Dunaliella, Chlorella, Scenedesmus*). Usage of a chamber filter press can attain a concentration factor of 245 for the relatively large size microalgae such as *C. proboscideum, which* produces a cake with 27% solids (Mohn, 1980). In addition a rotary drum vacuum or pressure filter can be used for filtration. A process using rotary drum pre-coat filtration was described for recovering microalgal biomass (Gudin and Therpenier, 1986). The application of cellulose fibers, sand filters for filtration of smaller diameter cells such as *Dunaliella* is not practical (Ben-Amotz and Avron, 1987). As an exception to these materials, *Dunaliella* can be filtered though diatomaceous earth. The cells could be recovered by passing diluted broth through diatomaceous earth. Thus the cells recovered are subjected to solvent extraction to recover β-carotene (Ruane 1977). The filtered alga was then directly extracted with organic solvent to recover β-carotene. As an alternative to conventional filtration for recover of biomass, membrane microfiltration and ultrafiltration can be used effectively for recovering microalgal biomass. Microfiltration can be used to recover delicate cells (Petrusevski et al., 1995). However it is restricted to small scale applications such as in aquaculture farms for recovery of microalgae which is used as a feed for shellfish larvae (Borowitzka, 1992). The possibility of using polymer membranes for continuous recovery of marine microalga Skeletonema costatum has been investigated (Rossignol et al., 1999). Ultrafiltration membranes made of polyacrylonitrile of 40-kDa molecular weight cutoff was found to be effective in recovering the cells (Rossignol et al., 1999). One disadvantage for micro-filtration membranes over ultra-filtration membranes is that the micro-filtration membranes though has greater initial fluxes than ultra filtration membranes, the former tends to clog more readily than the later. The major cost factors found with membrane filtration process are the need for membrane replacement and pumping. Generally, in handling of small volumes, microfiltration can be more cost-effective than centrifugation. However, for larger scale of production (e.g., >20 m^3 day^{-1}), centrifugation is better and cost effective in recovering biomass (MacKay and Salusbury, 1988).

3.3 BIOMASS PROCESSING

After the biomass is harvested by the techniques described in the previous section, it is processed further for the recovery of the target metabolites. These techniques are discussed below.

Dehydration

Removal of water in harvesting microalgal biomass results in concentration (50–200 folds) of microalgal biomass. Immediate processing of biomass is important to prevent from spoilage particularly in hot climate. The processing protocols depend on the product of interest. Dehydration or drying of biomass is most generally used to preserve biomass, if the biomass is the final product. Many drying methods have been reported for removing moisture from the microalgal biomass. Drying methods such as spray drying, drum drying, freeze-drying and sun drying are commonly employed in microalgal biomass preservation. Satisfactory results can be obtained if β-carotene rich *Dunaliella* is processed by spray drying, freeze-drying and drum drying. Good results were obtained in terms of uniformity of the algal biomass powder and stability of β-carotene in the biomass (Ben-Amotz and Avron, 1987). Spray drying however is specific for high value products (>$ 1000/ton), but the method can deteriorate some of the algal components like fatty acids and pigments. The expense of drying can be a significant obstacle to produce microalgal biomass powder for use in food and feeds. On the other hand, methods such as freeze-drying or lyophilization, commonly employed for drying microalgal biomass in laboratory scale seems expensive for large scale application. In addition, for some metabolites (fatty acids) solvent extraction of dried biomass is much better than extraction from undisrupted wet biomass. Freeze drying has been tried by some researchers e.g., freeze dried biomass is used for extraction of lipids from *Isochrysis galbana.*

Cell disruption

Most products from microalgae are intracellular in origin. In order to extract these products cell disruption is commonly used in downstream processing. (Ruane, 1977; Mendes-Pinto et al., 2001). Though cell disruption methods were developed initially for non photosynthetic cells they can be used for photosynthetic microalgal cells as well (Chisti and Moo-Young, 1986). Various disruption techniques have been explored for the release of intracellular metabolites from the microalgal cells. The technique used depends on various factors like the type of algal species, its susceptibility to cell disruption, the value of the product etc. In addition, each method has its *pros* and *cons.* Microalgal cells can be lysed by disruption process using glass and ceramic beads in bead mills. The cells get disrupted due to the abrasive action of the glass or ceramic beads on the cells. This method has been used to disrupt cells (Hedenskog et al., 1969) of *Scenedesmus obliquus, S. platensis* and *Monodus subterraneous.* High-pressure homogenization is another cell disruption method, which involves the passage of cell suspension through a rotor-stator assembly. The device is generally operated at very high pressures. The microalgal cells are lysed due to the high pressures and shear they encounter. This technique is commonly used to disrupt *Haematococcus* cells for astaxanthin recovery. Yet another technique is ultrasonication. Ultrasonication results in cavitation i.e. formation of small bubbles in the medium through which it is passed. These cavities or bubbles come in contact with the microalgal cells and are responsible for their disruption. Ultrasonication of microalgal suspensions are often used to disrupt small quantity of cells (Dunstan et al., 1992; Bermejo Roman et al., 2001). Acid, alkali or enzyme treatment is also commonly employed to lyse the microalgal cells. Here the cells get disrupted due to the action of the acid/alkali/ enzymes on the cell wall components. Though process involving treatment with alkali is an effective method for cell wall lysis, it is not suitable for sensitive products such as proteins. Nevertheless, alkaline lysis is used to isolate free fatty acids from microalgae. Other methods involve, grinding of dried biomass at low (–170°C) temperatures. This method has been used in *Haematococcus* to extract astaxanthin with butylated hydroxyl-toluene. However it is not practically applicable at large scale. Saponification of wet biomass directly with KOH-ethanol mixture has also been tried for extracting free fatty acids from *Porphyridium cruentum* (Gimenez et al., 1998) and *Phaeodactylum tricornutum* (Cartens et al., 1996; Ibanez Gonzalez et al., 1998).

Extraction and purification

Extraction is generally carried out using organic solvents like hexane, ethanol, chloroform and diethyl ether. High value metabolites such as astaxanthin, β-carotene and essential fatty acids (GLA) are extracted from algal biomass using solvent extraction. Solvents are used to extract essential fatty acids such as EPA, docosohexaenoic acid (DHA) and arachidonic acid (AA) from various microalgae. Fatty acid extraction by organic solvents from, *P.S. tricornutum, P. cruentum I. galbana, M. subterraneous* has been described in the literature.(Gimenez et al., 1998). Using buffers extraction of pigments such as phycobiliproteins from *P. cruentum* (Bermejo et al., 2001) and luetin from *C. vulgaris* have been reported. The extracts that are obtained from microalgae in crude form are usually filtered and purified by chromatographic techniques to arrive at pure metabolic product. Most recently supercritical fluid extraction has been used to recover astaxanthin (Lin et al., 2003), polyunsaturated fatty acids and other high value compounds from microalgae. Chromatographic techniques such as reverse phase, silica gel and arginated silica gel chromatography (Gimenez et al., 1998) have been used for recovering pure fatty acids. Protein purification is accomplished using ion exchange chromatography (Bermejo et al., 2001).

Economic Aspects

In any biotechnological process, the recovery of the product of interest is the critical factor, which governs the cost of the entire process. The economics of metabolite production from microalgae are influenced significantly by the cost of producing the biomass, the metabolite content in the biomass, and the cost of purification. Typically, the downstream recovery of microbial products accounts for 70–80% of the total cost of production (Chisti, 1998). For a product such as EPA, 60% of the costs arise from the recovery process. The biomass contributes 40% to the cost of EPA. This suggests that the microalgal biomass for producing EPA is significantly more expensive than the norms for bioprocesses. Clearly, monoseptic culture of microalgae in outdoor photobioreactors is expensive compared to heterotrophic growth of most commercial bacteria, yeast and fungi. Enhanced metabolite content in the alga in combination with better performing photobioreactors, can potentially lead to economically feasible processes for microalgal metabolites.

3.4 POTENTIAL MICROALGAL PRODUCTS

During recent years microalgae have been recognized as potential producers of some important high value metabolites. Some of these compounds are discussed below.

Nutracueticals

Microalgae are a source of many essential fatty acids. Indeed microalgae are considered as dietary food supplements to humans because of these essential fatty acids. Various unsaturated fatty acids include unique prostaglandins as source of essential unsaturated fatty acids (C18 and C22 fatty acids) which act as antecedents for many prostaglandins and ecosanoids. Various essential fatty acids from algal sources include linoleic-acid(C18:2) ã-linolenic acid(C18:3) (GLA), eicosapentaenoic acid (C20:5) (EPA), arachidonic acid (C20:4) (ARA) and decosahexanoic acid (C22:6) (DHA). Among these essential fatty acids GLA plays an important role and acts as a precursor of ARA and DHA. These in turn act as precursors for prostaglandins, prostaxylins, thromboxanes and leucotrienes, which are important products in the pharmaceutical industry. *Porphyridium cruentum* is known to accumulate nearly 40% of ARA (archidonic acid) in its lipids (Cohen, 1995). Extensive research shows that three of the key omega-3 fatty acids (ALA, EPA, and DHA) helps in reducing inflammation and prevent heart disease and arthritis. They are also known to assist in lowering the low-density lipids (LDL) and triglycerides. Currently these fatty acids are produced from fish oils. Both omega fatty acids (omega 6 and omega 3 fatty acids) are used as supplements in food. Spirulina is a potential source for gamma-linolenic acid (GLA). This fatty acid is an essential Omega-6 fatty acid found in the membrane bound lipids of the alga. A series of membrane bound proteins viz., NADPH-cytochrome b5 reductase, cytochrome b5 and a terminal desaturase are involved in the synthesis of these unsaturated fatty acids. Omega-6-fatty acid is a precursor for dihomo-ã-linolenic acid and arachidonic acids, these acids in turn, act as precursors for series 1 and 2 prostaglandins. Therapeutic value of this fatty acid has been established in treating various disorders in humans, which include rheumatoid arthritis, eczema, diabetes, multiple trauma and premenstrual syndrome. Though fungi are well-known source for PUFA (polyunsaturated fatty acid) production including GLA, the limitations in scale up undermine its large-scale cultivation. Plant sources have relatively less GLA content. Microalgae, *Spirulina platensis* is known to accumulate large quantity of GLA (31% TFA and 1–1.4% dry cell weight). *Spirulina platensis* is the organism of choice for large-scale production of omega-6 fatty acid. It can be targeted for GLA production owing to its low cost of cultivation, high content of GLA in total fatty acids and cheaper downstream processing protocols for biomass recovery. The fatty acid content in the alga can be easily enhanced by manipulating culture conditions like temperature, light-dark period. (Mahajan and Kamat, 1995) and the oil supplements be easily enhanced by manipulating culture conditions like temperature light-dark period. Spirulina can be an alternative source of GLA in future coming years.

Amino acids and Vitamins

Several amino acids could be potentially produced by microalgae. Some microalgae including *Spirulina platensis* accumulate high concentrations of proline when cultivated under high saline conditions. Excess production of proline is a measure of stress resistance against high salt concentrations (Ciferri, 1983). Apart from the amino acid proline, the blue green algae are known to have copolymers of aspartic acid and arginine (Ciferri, 1983) called cyanophycin bodies. Arginine is relatively a high value amino acid. Several vitamins are found in microalgae that are nutritionally important. Vitamins particularly antioxidant vitamin C, E and β-carotene are found in Microalgae. However, at present except β-carotene other vitamins are not produced commercially from microalgal sources.

Food Colours

Microalgae are potential sources of pigments; most of the pigments are photosynthetic accessory molecules except chlorophyll, which is a major photosynthetic pigment, found associated with the microalgae. Absorption of light at specific wavelengths is a characteristic feature of these pigments. Chlorophylls are green pigments, which contain a porphyrin ring. The porphyrin ring has the potential to gain or lose electrons; it is due to this process by which the chlorophyll molecule captures light during photosynthesis. Different types of chlorophylls exist in microalgae, the most dominating chlorophyll among the microalgae is chlorophyll a. This molecule finds its position in the major photosynthetic apparatus, which passes the energized electrons to the molecules that manufacture sugars. Major photosynthetic entities like plants cyanobacteria and other algal forms contain Chlorophyll a in large quantity when compared to other forms of Chlorophyll viz., Chlorophyll b and Chlorophyll c. The second category of Chlorophyll (Chl b) occurs in green algae and in plants. Chlorophyll c, however is restricted to the photosynthetic membranes of the group chromista and in dinoflagellates. In the food industry, the green color chlorophyll is used as a natural pigment ingredient in processed foods. Because of consumers growing interest on natural foods, chlorophyll is gaining importance as food additive. This is encouraging food processors to use natural colorants instead of artificial pigments. In the cosmetics industry, chlorophyll a is used in soaps and other cosmetics products.

Another class of pigments, commonly found in microalgal are carotenoids. Carotenoids are basically red, orange or yellow pigments. They include different types such as Carotenes, Xanthophylls, Leutin, Zeaxanthin etc. These compounds are primarily six carbon ring compounds connected by a chain of carbon atoms. They are not dissolved in water and hence they are found attached to the membranes within the cell. Carotenoids can not transfer energy (sunlight) to the photosynthetic pathway however they pass the absorbed energy to the major photosynthetic pigment, chlorophyll and thus these carotenoids are called photosynthetic accessory pigments. Fucoxanthin, the brown pigment is responsible for the color shown by kelps, other brown algae and diatoms. Similarly astaxanthin the red colored secondary carotenoid is responsible for the red color found in the encysted state of the Chlorphyte *Haematococcus pluvialis*. The value of this carotenoid is approximately $3000 per kg. Astaxanthin finds major applications in aquaculture where it is used as feed for salmon. According to a recent market report (www.bccresearch.com) the current worldwide market value of all commercially-used carotenoids is estimated at $ 887 million for 2004 and expected to rise at an average annual growth rate of 2.9% to just over $ 1 billion. *Haematococcus pluvialis* accumulate nearly 5% of astaxanthin in its cell dry biomass. In spite of its high carotenoid content hardly few commercial companies are exploiting this alga for astaxanthin production, and the reason can be possibly due to the lack of appropriate technology for cultivating this alga for astaxanthin production. Two companies were engaged actively in production of astaxanthin from *Haematococcus pluvialis*, both based at Hawaii, Aqua search Inc, and Cyanotech. They are developing a two stage process technology for production of astaxanthin .The vegetative phase is carried out in tubular systems (enclosed reactors) and the astaxanthin induction phase is carried out in open ponds.

The third class, phycobilins are water-soluble pigments, and hence they are most commonly found in either cytoplasm or in the stroma of the chloroplast. They occur in large quantity in Cyanobacteria (blue green algae) and in Rhodophyta (red algae) . The phycobilins can be resolved into two classes of compounds, the blue phycobilins found in cyanobacteria are called phycocyanins and the red phycobilins found in rhodophyta are called phycoerythrins. In fact the nomenclature of these algae was based on the presence of the type of pigment. Phycobilins functionally absorb energy in the regions of the visible spectrum that are less utilized by chlorophyll, through fluorescence; they convey the energy to chlorophyll molecule at the photosynthetic reaction centre. The blue green algae *Spirulina platensis* is a potential source for the production of phycocyanin for commercial purpose. Natural colourants, such as Phycocyanin, are gaining importance over synthetic colours, since they are non carcinogenic and non toxic. The applications of phycocyanin in foods include colouring of milk products, chewing gums, ice creams, soft drinks etc. Phycocyanins are also suitable for use as colourants in cosmetics. In addition the pigment finds application as fluorescent dyes for flow cytometry and in immunological assays.

Speciality Products

Amongst microalgae the toxic microalgae are potential sources of antifungal, antibacterial antiviral and anticancer drugs. To date industry has not exploited microalgae for new pharmaceuticals, mainly for economic reasons. Blue-green algae are excellent sources of new cytotoxins, fungicides, and viricides such as Anatoxins, Isonitriles and Paracyclophanes. In diagnostics, the unique properties of the pigments (fluorescence) have been used as a tool in diagnosing the antibodies. Particularly the phycobiliproteins of the blue green and red micro algae are used as fluorescent labels in both research and diagnostic kits. The pigments, pycocyanin and phycoerythrin show fluorescence at particular wavelength. Fluorescence is the phenomenon in which absorption of light of a given wavelength by the pigments (fluorescence) is followed by the emission of light at longer wavelengths. The phycobilins because of these unique fluorescence properties can be used as chemical tags of antibodies. This technique is known to be extremely useful in cancer research for tagging cancer cells. However, for diagnostic applications very little biomass in the order of few tens of kilograms per year are required hence the markets for these products are minimum in terms of biomass requirement. Another important application of microalgal derived compounds in diagnostics is production of isotopically labelled compounds. This can be accomplished by growing micro algae in closed photo bioreactors on $^{13}CO_2$ or in D_2O which allows the production of isotopically labelled compounds. However the requirement of biomass for producing such labelled compounds is again minimum. However, the value of such labelled compounds is of several million dollars and closed photobioreactors will suffice the production of such labeled compounds from microalgal biomass.

Miscellaneous

Currently the sulfated polysaccharides from sea weeds (macroalgae) are commercially used as thickening and flocculating agents in food industry. Some of the polysaccharides such as carrageenans are produced from both microalgae and macroalgae (seaweeds). They are used as thermo reversible and thickening agents. Carrageenans from red microalgae, such as *Porphyridium cruentum* and *P. aerugineum*, have similar properties of viscosities and gelling behavior when compared to macroalgae. The polysaccharides from microalgae are uniform and have more reliable properties than macroalgae which show significant variation of polysaccharide production from batch to batch. The polysaccharide content depends on environmental factors, light intensity and nitrogen supply. In spite of the product significance there is no commercial process developed for production of these polysaccharides from the microalgae. The high cost production of these products from microalgae and the possibility of low cost production from macroalgae (seaweeds) could be some reasons as to why the microalgae are not being properly exploited. Apart from carrageenans, other polysaccharides from microalgal sources include, laminarin; starch; inulin; fucoidin and polymers comprised of xylose, arabinose, rhamnose and glucuronic acids. Flocculation using biological entities (Bio-flocculants) by algae has potential industrial applications in water clarification. The flocculant produced by cyanobacterium *Phormidium* is known to cause clay sedimentation in drainage troughs in Israel. However, the commercial potential has not been demonstrated for these products.

Chapter 4

BIOSORPTION

4.1 INTRODUCTION

The discharge of heavy metals into aquatic ecosystems has become a matter of concern in India over the last few decades. These pollutants are introduced into the aquatic systems significantly as a result of various industrial operations. Heavy metals are usually classified as the following three categories: toxic metals (such as Hg, Cr, Pb, Zn, Cu, Ni, Cd, As, Co, Sn, etc.), precious metals (such as Pd, Pt, Ag, Au, etc.) an radionuclides such as U, Th, Ra, Am, etc.), whose specific weight is usually more than 5.0 g/cm^3. The toxicity of heavy metals occurs even in low concentration of about 1.0–10 mg/l. Some strong toxic metal ions, such as Hg and Cd, are very toxic even in lower concentration of 0.001–0.1 mg/l.

Some of the industries, which are responsible for heavy metal pollution, are given in Table 4.1.

Table 4.1 Heavy metal pollution form industrial activites

Heavy metal	Industry
Cr (VI)	Tannery waste, colors and pigments, metal platting, batteries, paper and pulp industries
Cd (II)	Ni-Cd batteries, pigments, pesticide, electroplating waste, mine waste etc.
Zn	Mining industry, galvanizing plants, manufacturing brass and other alloys etc.
Cu	Metallurgical industry, process industry
Pb	Paints, lead water pipes, gasoline containing tetraethyl lead additives etc.
Ni	Nonferrous metal mineral processing, paint formulation, electroplating, copper sulfate manufacturing and steam electric power plants etc.

Heavy metal toxicity has become a major concern today due to the deleterious effects they can cause. For e.g. Chromium (VI) is carcinogenic and can modify the DNA transcription process. It is also reported to cause epigastric pain, nausea, vomiting, severe diarrhea, and hemorrhage (Dakiky et al., 2002). Similarly other heavy metals are also harmful to human health. Hence, it is very important to treat these effluents containing heavy metals. Conventional methods like ion exchange and precipitation are generally employed to remove the same. These are discussed in brief.

Conventional Methods

- **Reverse Osmosis:** It is a process in which heavy metals are separated by a semi-permeable membrane at a pressure greater than osmotic pressure caused by the dissolved solids in wastewater. The disadvantage of this method is that it is expensive.
- **Electrodialysis:** In this process, the ionic components (heavy metals) are separated through the use of semi-permeable ion selective membranes. Application of an electrical potential between the two electrodes causes a migration of cations and anions towards respective electrodes. Because of the alternate spacing of cation and anion permeable membranes, cells of concentrated and dilute salts are formed. The disadvantage is the formation of metal hydroxides, which clog the membrane.
- **Ultrafiltration:** They are pressure driven membrane operations that use porous membranes for the removal of heavy metals. The main disadvantage of this process is the generation of sludge.
- **Ion-exchange:** In this process, metal ions from dilute solutions are exchanged with ions held by electrostatic forces on the exchange resin. The disadvantages include: high cost and partial removal of certain ions.
- **Chemical Precipitation:** Precipitation of metals is achieved by the addition of coagulants such as alum, lime, iron salts and other organic polymers. The large amount of sludge containing toxic compounds produced during the process is the main disadvantage.

Various disadvantages like incomplete metal removal, high reagent and energy requirements, generation of toxic sludge or other waste products that require careful disposal has made it imperative for a cost-effective treatment method that is capable of removing heavy metals from aqueous effluents. Biosorption is gaining considerable interest as a potential alternative method to the conventional techniques.

4.2 WHAT IS BIOSORPTION?

Biosorption can be defined as the ability of biological materials to accumulate heavy metals from wastewater through metabolically mediated or physico-chemical pathways of uptake (Fourest and Roux, 1994). The biosorption process involves a solid phase (sorbent or biosorbent; biological material) and a liquid phase (solvent, normally water) containing a dissolved species to be sorbed (sorbate, metal ions). Due to higher affinity of the sorbent for the sorbate species, the latter is attracted and bound there by different mechanisms. The process continues till equilibrium is established between the amount of solid-bound sorbate species and its portion remaining in the solution.

The major advantages of biosorption over conventional treatment methods include (Nortan et al., 2004):

1. Low cost
2. High efficiency
3. Minimization of chemical and/or biological sludge
4. No additional nutrient requirement
5. Regeneration of biosorbent
6. Possibility of metal recovery, could be employed most effectively in a concentration range below 100 mg l^{-1}.

Biosorbents

Algae, bacteria, fungi and yeasts have proved to be potential metal biosorbents (Volesky, 1995). Biosorbents represent 'chemicals' capable of sequestering a relatively large amount of the metal. Strong biosorbent behavior of certain types of microbial cells towards metallic ions is a function of the chemical make up of the microbial cells. Some types of biosorbents could be broad range, binding and collecting the majority of heavy metals with no specific priority, while others can even be specific for certain types of metals. There are potent biosorbent materials among easily available biomass types from all three groups: algae, fungi and bacteria, the former two perhaps giving broader choices (Volesky, 1995).

Although many biological materials bind heavy metals, only those with sufficiently high metal-binding capacity and selectivity for heavy metals are suitable for use in a full-scale biosorption process. The first major challenge for the biosorption field was to select the most promising types of biomass from an extremely large pool of readily available and inexpensive biomaterials. Although this task is not complete, a large number of biomass types have been tested for their metal binding capability under various conditions. Table 4.2 lists some of the species that possess a metal-binding capacity comparable with commercial synthetic cation-exchange resins.

Table 4.2 Metal-binding capacity of some biosorbents

Biomass type	Biosorbent capacity (meq g^{-1})[a]	Reference
Ascophyllum sp.	2–2.5	Chong and Volesky, 1995
Eclonia radiate	1.8–2.4	Sharma and Forster, 1993
Rhizopus arrhizus	1.1	Fourest and Roux, 1994
Sargassum sp.	2–2.3	Fourest and Volesky, 1996
Commercial resins	0.35–5.0	Kratochvil and Volesky, 1998
Peat moss	4.5–5.0	Jeffers *et al.*, 1991

[a] meq g^{-1} (milliequivalents per gram of dry biomass) is a concentration term corrected for ion or active-site valence.

Although several proprietary biosorption processes (such as AlgaSORB™; Darnall, 1991 and AMT-Bioclaim™; Brierley et al., 1986) were developed and commercialized in the early 1990s, a lack of understanding of the mechanism underlying the metal-sorption process has hindered adequate assessment of process performance and limitations, and thus the expected widespread application of biosorption. The next real challenge for the field of biosorption was to identify the mechanism of metal uptake by dead biomass.

Mechanism

In the process of biosorption, heavy metal ions are primarily accumulated on the surface of the cell. Once the metal ion has diffused to the cell wall, it will bind to sites on the cell surface, which exhibit chemical affinity for the metal. This process may contain number of processes such as adsorption, ion exchange, coordination, complexation, chelation, and microprecipitation (Dönmez et al., 1999). Several possible mechanisms for metal biosorption have been scrutinized and several systematic studies of the binding of heavy metals to biomass, particularly fungal and algal biomass, have been undertaken. Although not all of the potentially applicable biosorbents have yet been systematically examined, a substantial body of evidence has been collected that identifies ion exchange as the principal mechanism of metal biosorption. Moreover, strong links have been established between metal sequestration by various types of biomass and metal binding by biopolymers extracted from the cells of the corresponding organisms (Macaskie and Dean, 1990). The interaction with cell walls or with membranes is the initial process in any biotic action. The exterior surfaces have a common composition of proteins and carbohydrates with which the metallic species could react. Metal ions could displace other metal ions or protons, and a reversible pH titration indicated the existence of labile protons whose loss did not lead to substantial structural changes (Crist et al., 1981). Though the cell wall composition for all algae is unknown, protein content of the cell wall is reported to be 16-27% (Siegal and Siegal, 1973). Amino acids in the proteins could also provide such functional groups. The polysaccharides of the cell wall could also provide the amino and carboxyl groups as well as the sulfate. The amino and carboxyl groups, the imidazole of histidine, and the nitrogen and oxygen of peptide bonds could be available for characteristic coordination bonding with metallic ions like Cu^{2+} Bond formation could be accompanied by displacement of protons dependent in part on the extent of protonation as determined by the pH. Metallic ions could also be electrostatically bonded to unprotonated carboxyl oxygen and sulfate.

4.3 FACTORS AFFECTING BIOSORPTION

The following factors affect the biosorption process:

pH

For biosorption of heavy metal ions, pH is one of the most important environmental factors. The pH value of solution strongly influences not only the site dissociation of the biomass' surface, but also the solution chemistry of the heavy metals: hydrolysis, complexation by organic and/or inorganic ligands, redox reactions, precipitation, the speciation and the biosorption availability of the heavy metals (Esposito et al., 2002). The biosorptive capacity of metal cations increases with increasing pH of the sorption system, but not in a linear relationship. On the other hand, too high pH value can cause precipitation of metal complexes, so it should be avoided during experiments. For different biosorption system of metal ions, the optimal pH is different. However, metal anions, such as chromium(VI), exhibit different pH features from metal cations. Generally speaking, a low pH value is favorable for biosorption. The effect of pH on biosorption of Cr(VI) has been investigated by various researchers using a variety of different types of biomass. Optimal biosorptive removal of Cr(VI) at low pH (2.0) has been reported for *Rhizopus nigricans, Bacillus* sp. and *Dunaliella* sp. It is supposed that anions could be sorbed on the protonated active sites of the biosorbent primarily electrostatically in nature. At very low pH values, the surface of sorbent would also be surrounded by the hydronium ions, which enhance the chromium(VI) interaction with binding sites of the biosorbent by greater attractive forces. As the pH increases, the overall surface charge on the cells became negative and decreases the biosorption capacity.

Temperature

Temperature also has an influence on the biosorption of metal ions, but to a limited extent under a certain range of temperature, which indicates that ion exchange mechanism exists in biosorption to some extent. Biosorption process is usually not operated at high temperature because it will increase the operational cost (Wang, 2002). Adsorption reactions are normally exothermic, so biosorption capacity increases with decrease of temperature (Kapoor and Viraraghavan, 1997). The decrease of biosorption capacity at higher temperature may be due to the damage of active binding sites in the biomass.

Contact time

Biosorption is a rapid process and generally equilibrium is reached very fast. However, in practice, it is necessary to optimize the contact time, considering the efficiency of desorption and regeneration of the biomass. Ferraz et al., (2004) optimised the sorption time for Cr(III) by S. *cerevisiae* from a brewery company in the sorption–desorption process, the result showed that a 30 min sorption period was the best option to ensure the metal removal from solution and good recovery from biosorbent.

Competing ions/co-ions

Real industrial effluent usually contains various ionic components, including metal cations and anions. Some studies indicated that cations and anions additional to the ions of interest have a generally detrimental impact on metal accumulation (Suh and Kim, 2000). Usually, the biosorption capacity of one metal ion is interfered and reduced by co-ions, including other metal ions and anions present in solution. However, the gross uptake capacity of all metals in solutions remains almost unchangeable. The decrease of metal uptake in competitive conditions was thought to be a response to increased competition between same charged species for binding sites. The presence of anions also affects the biosorption of metal ions. Kapoor and Viraraghavan (1997) reported that the biosorption capacity decreased in the presence of ethylenediamine tetraacetate (EDTA), sulphate, chloride, phosphate, carbonate, glutamate, citrate and pyrophosphate. The anions existing in the solution may form a complex with the metal ions, which would reduce the metal biosorption capacity seriously.

Initial concentration of metal ions and biomass

The uptake rate of the metal ion will increase along with increasing the initial concentration if the amount of biomass is kept unchanged. Contrary to that, biosorptive capacity of the metal ions is inversely proportional to the initial concentration of the biomass when the initial concentration of metal ions is kept constant. Increase of the biomass concentration of the biosorption system could result in increasing the sorption site interactions. When the biomass concentration is low, metal ions in the solution would not only be adsorbed to the surface of the biomass, but also enter into intracellular part (Wang, 2002). Biosorptive capacity of metal ions was reported to be related to the ratio of the concentration of initial metal ions to the concentration of biomass. Both aspects cannot be neglected when assessing the influence of concentration of the metal ion and biomass on biosorption (Schiewer and Volesky, 1995a).

4.4 EVALUATION OF SORPTION PERFORMANCE

The assessment of a solid–liquid sorption system is usually based on two types of investigations namely, equilibrium batch sorption tests and dynamic continuous-flow sorption studies (Volesky and Holan, 1995). Equilibrium isotherm models are usually classified into empirical equations and mechanistic models, based on the mechanism of metal ion biosorption. Mechanistic models can be used not only to represent, but also to explain and predict the experimental behavior (Liu et al., 2002b; Pagnanelli et al., 2002). The empirical models for single solute systems used to describe the biosorption equilibrium are Langmuir, Freundlich and Brunauer–Emmett–Teller (BET) models. Langmuir model (L type, based on monolayer adsorption of solute) and Freundlich model (F type, developed for heterogeneous surfaces) are the most widely accepted and used in a number of references. The BET model describes the multi-layer adsorption at the adsorbent surface, and assumes that a Langmuir isotherm applies to each layer. These models can provide information of metal-uptake capacities and differences in metal uptake between various species (Pagnanelli et al., 2002; Volesky and Holan, 1995). Scatchard plots, which originally describe the attractions of proteins for small molecules and ions, were developed to describe the biosorption equilibrium. The total number of binding sites may be extrapolated from Scatchard plots if complete saturation of the biomass binding sites is not observed (Kapoor and Viraraghavan, 1995). To describe two or multi-metal ion biosorption systems, various extended Langmuir models (also called comparative Langmuir model) or Freundlich type models have been developed. More details or examples on those empirical models can be obtained from the relevant references (Aksu et al., 1997; Kapoor and Viraraghavan, 1995; Pagnanelli et al., 2002; Volesky and Holan, 1995). However, those empirical models can hardly explain the biosorption behaviour with a meaningful physical interpretation, and predictive conclusions cannot be drawn for systems operating under different conditions. Mechanistic conclusions from the good fit of models alone, should be avoided. Moreover, biosorption isotherms may exhibit an irregular pattern due to the complex nature of both the biomaterials and its varied multiple active sites, as well as the complex solution chemistry of some metallic compounds (Volesky and Holan, 1995; Kapoor and Viraraghavan, 1995). Therefore, mechanical models have been proposed not only to represent experimental behaviour but also to explain and predict biosorption behaviour based on the understanding of biosorption mechanisms. Two kinds of mechanistic models are ion exchange model and surface complexation model. (Figueira et al., 2000; Schiewer and Volesky, 1996, 1997; Volesky, 1993; Yang and Volesky, 1999). Kinetics studies and dynamic continuous-flow investigations, offering information on the rate of metal uptake, together with the hydrodynamic parameters, are very important for biosorption process design (Volesky and Holan, 1995). Many models have been described in Chapter 6. The biosorption studies carried out on different microalgae has been complied from recent literature and reported in Table 4.3.

Table 4.3 Literature reports on biosorption

Biosorbent	Metal ion	pH	Temperature	q_{eq}	Adsorption isotherm	Kinetics	Reference
Chlorella vulgaris	Ni^{2+}	4.5	15-45°C (max at 45°C)	60.2 mg g^{-1} (45°C)	Freundlich, Langmuir, Redlich-Peterson	Saturation type kinetic model	Aksu (2002)
Chlorella vulgaris	Cu^{2+}	3.5	35°C	60 µg . metal ion mg $protein^{-1}$		Michaeliss–Menten	Mallick (2002)
Chlorella vulgaris	Ni^{2+}	3.5, 6.5	35°C	53 µg metal ion mg $protein^{-1}$		Michaelis–Menten	Mallick (2002)
Chlorella vulgaris	Cu^{2+}	5.0	25°C	40 mg g^{-1}	Freundlich, Langmuir		Dönmez et al., (1999)
Chlorella vulgaris	Ni^{2+}	4.5	25°C	42.3 mg g^{-1}	Freundlich, Langmuir		Dönmez et al., (1999)
Chlorella vulgaris	Cr^{6+}	2.0	25°C	23 mg g^{-1}	Freundlich, Langmuir		Dönmez et al., (1999)
Spirulina sp.	Cr^{3+}	7	35°C	10 meq g^{-1}	Langmuir	First order kinetics	Chojnacka et al., (2005)
Spirulina sp.	Cd^{2+}	7	35°C	2.83 meq g^{-1}	Langmuir	First order kinetics	Chojnacka et al., (2005)
Spirulina sp.	Cu^{2+}	7	35°C	6.17 meq g^{-1}	Langmuir order	First order kinetics	Chojnacka et al., *(2005)*
Scened-esmus obliquus	Cu^{2+}	5.0	25°C	20 mg g^{-1}	Freundlich, Langmuir		Dönmez et al., (1999)
Synechoc-ystis sp.	Cu^{2+}	5.0	25°C	23.4 mg g^{-1}	Freundlich, Langmuir		Dönmez et al., (1999)
Scened-esmus obliquus	Ni^{2+}	4.5	25°C	18.7 mg g^{-1}	Freundlich, Langmuir		Dönmez et al., (1999)
Scenedesm-us obliquus	Cr^{6+}	2.0	25°C	15.6 mg g^{-1}	Freundlich, Langmuir		Dönmez et al., (1999)
Synech-ocystis sp.	Cr^{6+}	2.0	25°C	19.2 mg g^{-1}	Freundlich, Langmuir		Dönmez et al., (1999)

4.5 IMMOBILIZATION

Enzymes and cells can be immobilized by binding to a suitable carrier or by entrapment. Binding may be physical adsorption, ionic or covalent (chemical) binding. A combination of methods can also be used, for example, adsorption and subsequent cross-linking with bifunctional reagents such as glutaraldehyde. Entrapment is possible in polymer gels, fibers, and in microcapsules. Entrapment is usually carried out in alginate or k-carrageenan (Onaka et al., 1985). However, the major drawback of entrapment is the mass transfer limitations even when using very small alginate beads (Onaka *et al.* 1985). Therefore entrapment has generally been replaced by adsorption or ionic binding on various carriers. The most important advantage of entrapment is the relatively low price of the polymer gel. Cell immobilization on a solid carrier is carried out by physical adsorption due to electrostatic forces or by covalent binding between the cell membrane and the carrier. Systems using immobilized cells on a surface are popular due to the relative ease of carrying out this type of immobilization. The strength with which the cells are bonded to the carrier as well as depth of the biofilm often varies and is not readily determined. As there are no barriers between the cells and the solution, cell detachment and relocation is possible with potential establishment of equilibrium between adsorbed and freely suspended cells. Examples of solid carriers used in this type of immobilization are cellulosic materials (DEAE-cellulose, wood, sawdust, delignified sawdust), inorganic materials (polygorskite, montmorilonite, hydromica, porous porcelain, porous glass), etc. Solid materials like glass or cellulose can also be treated with polycations, chitosan or other chemicals (pre-formed carriers) to enhance their adsorptionability (Norton, 1994). The following properties should be considered when choosing the carrier:

1. Price
2. Convenience of immobilization
3. Type of immobilization
4. Cell load
5. Mass transfer
6. Channeling or blocking of contactors
7. Stability
8. Rigidity
9. Regeneration
10. Possibility to use various contactor designs
11. Possibility of fluidization (also for regeneration).

Many of these points may alone prevent the use of a carrier, for example, a high price. Some of them are interdependent: if the price is very low, there is not necessarily need for any regeneration. Several types of carriers are available at present, for example, cellulose derivates, porous glass, silicon rubber, diatomaceous earth, silicon carbide, and wood chips, to name some.

4.6 CONTINUOUS OPERATIONS-PACKED BED COLUMN

The biosorption of heavy metals by dead microalgal biomass is a highly cost-effective new alternative for the decontamination of metal-containing effluents. The understanding of the mechanisms of metal biosorption will allow the process to be scaled up and used in field applications. Packed-bed sorption columns is perhaps most efficient for this purpose. Regenerating the biosorbents increases the process economy by allowing their reuse in multiple sorption cycles. The process results in metal-free effluents and small volumes of solutions containing concentrated metals, which can be easily recovered. Recent studies of biosorption phenomena have revealed that biosorbents, very much like synthetic ionexchange resins, can be prepared in different ionic forms, such as protonated (H-form) or saturated with Ca^{2+}, Mg^{2+}, Na^{+} etc., by washing the biomass with mineral acids, salts and/or bases. Consequently, questions arise as to which ionic form should be used for removing heavy metals and which chemicals should be selected for desorbing the metals in order to assure optimal performance of the biosorption columns. The overall performance of flow-through columns is strongly related to the length and shape of the ion exchange zone that develops during sorption and regeneration (Fig. 5.1.) When the metal ion 'breaks through' and becomes detectable at the column exit at a preset (or given) concentration, the useful life of the column is over (the column service time). However, at that time the column is still not completely used up (saturated) because of the unsaturated transfer zone inside, which should be minimized. This zone develops between the section of the column that is saturated with heavy metal ion and the section that still contains fresh biosorbent. As the loading or regeneration of the biosorbent progresses, the zone moves along the column in the direction of the liquid flow (usually downwards). When the sorbent saturation zone approaches the end of the column, the metal ion concentration in the outlet stream increases sharply and the useful life of the column is over—this is the breakthrough point, and the time before breakthrough occurs is the service time of the column. Clearly, the shorter the dynamic ion-exchange zones in the column, the longer the column service time and the larger the fully utilized sorbent portion inside the column. Correspondingly and conversely for the saturated-column-desorption cycle: it will be more efficient when the reverse ion-exchange zone during the column-regeneration stage is short, reflected in a steep breakthrough curve. In this case, the deposited metal comes out of the column at a very high concentration in a small volume of the elution liquid.

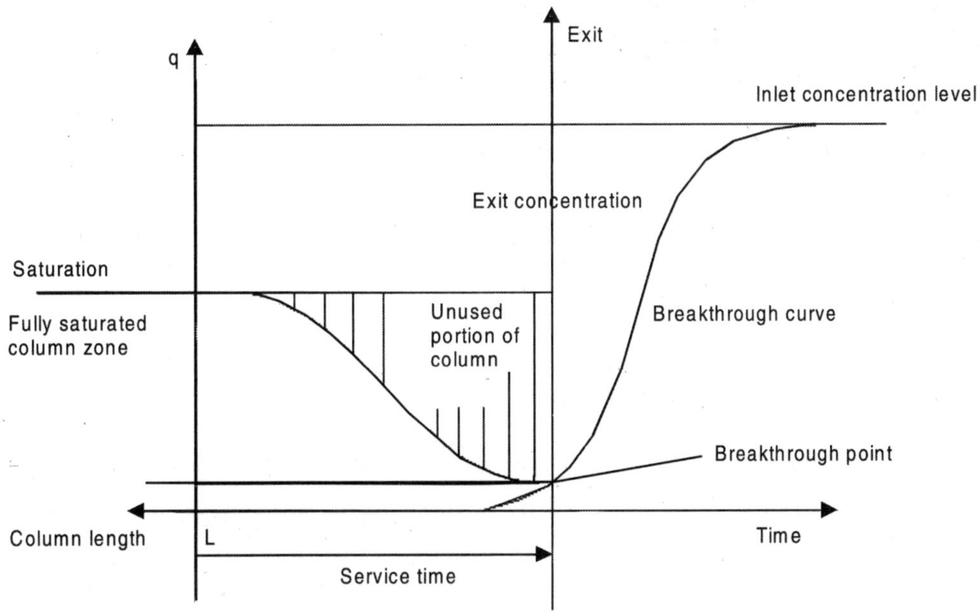

Fig. 4.1. *Flow-through a packed-bed biosorption column.*

A very important practical consideration arises from the affinity of different sorbates for the sorbent material. For the case of a spècies A sorbing onto B-saturated biosorbent, the theory of ion exchange distinguishes between two different scenarios of column performance, depending on the respective affinities of A and B for the sorbent material. These two scenarios are reflected in the shapes of the breakthrough curves resulting from such column operation (Fig. 5.2). If species A is more strongly bound to the resin than species B (i.e. affinity of A > affinity of B), then the zone that develops in a column is short and retains its shape as it moves down the column; however, if the affinity of B is greater than that of A, the zone extends across a large section of the column and is prone to further broadening as it travels along the column during the operation time. This indicates that a high degree of biosorbent utilization or regeneration is achieved only if the species sorbing onto a biosorbent has a higher affinity than the one used for presaturating the sorbent material. Therefore, the selection of both the ionic form of the biosorbent for the loading stage and the regenerant for the regeneration stage should assure that at least one, but ideally both, of these stages would follow the pattern of strongly binding of A replacing weakly bound B. Since, biosorption is a process during which toxic heavy metals may be exchanged for nontoxic species, the relative positions of protons, heavy metals and light metals in the affinity series determine the suitability of the ionic forms and regenerants. Unfortunately, at the present time, most of the affinity series reported in the literature for different biosorbents include only heavy metals. One of the priorities of future research should therefore be to establish complete affinity series for selected biosorbents, including heavy metals, light metals and protons.

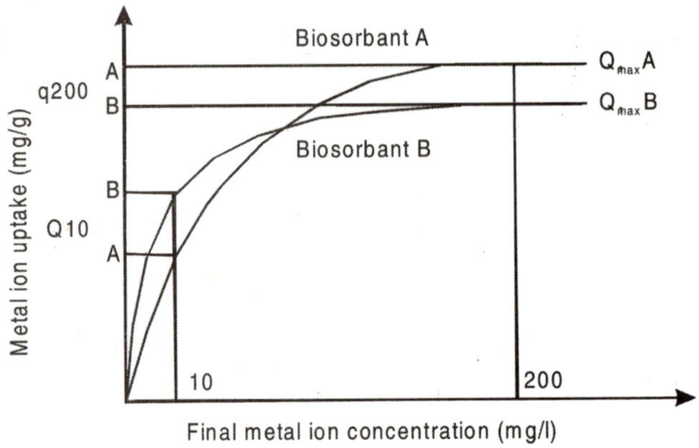

Fig. 4.2. *Breakthrough curves obtained from operation of biosorption columns.*

An unfavourable breakthrough curve is flat and trailing, indicating a wasteful long transfer zone inside the column. A favourable breakthrough curve, on the other hand, is steep and sharp, showing the effective utilization of the biosorbent material inside the column.

The feasibility and efficiency of a biosorption process depends not only on the properties of the biosorbents, but also on the composition of the waste-water. The faith in emerging technologies in waste-water treatment too often hinges on the success of pilot tests carried out with a specific industrial effluent. It is particularly important that only effluents that can be treated efficiently are selected for extended testing of the biosorption process. The majority of industrial effluents contain more than one toxic heavy metal. Consequently, biosorption in columns involves competitive ion exchange, in which several toxic heavy metals compete for a limited number of binding sites. In practical applications, the loading of biosorption columns has to be stopped as soon as the concentration of one of the toxic species in the column effluent exceeds the regulatory limit. It is well known from the theory and practice of ion exchange that the species that flows out of the column first is the one with the lowest affinity for the resin. Hence, the service time of biosorption columns is determined by the toxic metal in the feed that has the lowest affinity. The service time of biosorption columns can be reduced if the industrial effluent contains a nontoxic species interfering with the sorption of the toxic metals. The variety of ionic species of iron is a particular concern for biosorption processes treating mining effluents (e.g. acidic mine drainage). Although the cations of Fe^{2+} can potentially compete with cations of toxic metals for the binding sites in the biomass, Fe^{3+} is usually present in the form of suspended solids. This is the reason why careful evaluation of the applicability of a biosorption process for treating mining effluents is required (Kratochvil and Volesky, 1998). The selectivity of biosorbents for toxic metals over Fe^{2+} should be established (Kratochvil and Volesky, 1998). Furthermore, the effect of the level of suspended solids in the feed on column performance should be examined. Table 4.4 gives some of the work done in past few years in column studies for biosorption.

Table 4.4 Literature reports on column studies

Biosorbent	Immobiliation method	Metal ion	Parameters	Model	Reference
Chlorella vulgaris	Entrapment in Ca-alginate gel beads and agarose	Cu^{2+}	Flow rate, inlet metal ion concentration.		Aksu et al., 1997
Sargassum wightii	—	Cu^{2+}	Bed height, flow rate and inlet solute concentration	Thomas, Yoon–Nelson and modified dose–response, BDST model	Vijayaraghavan and Prabu, 2006
Turbinaria ornata	—	Cu^{2+}	Bed height, flow rate	Thomas model, BDST model	Vijayaraghavan et al., (2005)
Chlorella Sorokiniana:	Loofa sponge-immobilized	Ni^{2+}			Akhtar et al., (2004)
Spirogyras sp.	Silica-immobilised	Cu^{2+}	Flow rate, pH, equilibration time and other variables		Singh and Prasad (2000)
Chlorella vulgaris		Cu^{2+}, Cd^{2+}, Fe^{3+}, Sn^{4+}	Effect of solvents on metal ion adsorption		Al-Qunaibit et al., (2005)
Ulva reticulata		Cu^{2+}, Ni^{2+}, Co^{2+}	Bed height and Flow rate	Thomas model, BDST model	Vijayaraghavan et al., (2005)
Sargassum		Cu^{2+}, Cd^{2+}, Zn^{2+}, Fe^{3+}	Effects of competitive ion exchange, concentration overshoots	Ion exchange equilibrium column model	Kratochvil and Volesky (1997)
Sargassum fluitans		Cu^{2+}	Effects of Ca-, H- and H/Ca-*cycles*	Ion exchange model	Kratochvil and Volesky (1997)

4.7 BIOSORPTION COUPLED WITH METABOLITE SEPARATION

One of the major drawbacks of biosorption, which often limits its use for large scale removal of heavy metals, is the requirement of large amount of biomass. As discussed in the earlier chapters, the cost of cultivating these microalgae solely for the purpose of biosorption of heavy metals can be huge and impractical. On the other hand there are a number of industries, which are cultivating microalgae for recovery of valuable metabolites like asthaxanthin and β-carotene. These processes would in turn result in the generation of large amount of biomass, which is often discarded. It would be an economical alternative to use such spent microalgae for biosorption. This concept has not yet been exploited and research in this area is still in its infancy. The authors have proposed a two step process, which describes the extraction of a high value metabolite from algae and subsequent biosorption on the spent algae. Such economical strategies can be easily adopted on a large scale. Yet another reason for coupling metabolite extraction with biosorption can be justified as follows. It is well established that the rate limiting step in biosorption is the penetration of the metal ions through the cell wall. If the cell wall is permeabilized, mass transfer resistance for the diffusion of metal ions will be decreased and biosorption rate can be increased. Very often biosorption efficiency is increased by the addition of organic solvents. This has been reported by Al-Qunaibit et al., (2005) where permeabilization of the algal cells was achieved by addition of solvent in the heavy metal ion solution. However, addition of solvents increases the cost of the process. The algal cells can be permeabilized for the extraction of metabolites using a suitable techniques like solvent extraction (Pisal and Lele, 2005) or shear (Khirsagar et al., 2002). Subsequent biosorption of heavy metals on the spent algae, not only serves as value addition to the process but also makes it more economical. Preliminary experiments to demonstrate the efficacy of spent algae have been carried out by the authors. In their study, β-carotene was extracted from *Spirulina plantensis* and spent *Spirulina* biomass was used for biosorption of Chromium ions. The spent algal biomass showed better efficiency as compared to the control biomass (from which pigments are not extracted). It is hypothesized that the organic solvent used for extraction of β-carotene results in the rupture of the cell membrane of spirulina plantensis, which causes in an increase in the surface area available for biosorption.

Chapter 5

GENETIC MODIFICATION

5.1 INTRODUCTION

Metabolic engineering has emerged in the past decade as an interdisciplinary field aiming to improve cellular properties by using modern genetic tools to modify pathways. An impressive portfolio of metabolic engineering applications has resulted from this activity. On the experimental side, applications have focused on pathway modifications in (primarily) prokaryotic and eukaryotic systems, and the development of new tools for the introduction of genetic controls and assessment of the resulting cell physiology. The field is poised to benefit further from developments in genomics, proteomics and new technologies, such as DNA microarrays, that probe the expression phenotype of cells. Many industrially relevant problems in biotechnology are well addressed within a metabolic engineering framework that identifies unknown bioreaction pathways and elucidates the metabolic regulation of pathways as well as reaction and/or transport bottlenecks. Genetic Engineering is defined as the introduction of new or amplified genetic information into an organism by gene cloning and recombinant DNA methods. Genetic manipulation encompasses all techniques that can be used to improve the characteristics of organisms by modifying their genetic makeup. Thus, it includes the modification of an organism's genome by mutagenesis and selection as well as by genetic engineering. Although eukaryotic algae are of value both as organisms for fundamental biological research and as a resource for the biotechnology industry, their importance in these areas would be enhanced significantly by the availability of a genetic engineering methodology. Algae are used as model systems to study many different aspects of cell biology, biochemistry, and physiology. Major advances can be made in these studies through the use of recombinant DNA techniques to manipulate individual genes within the organism or to introduce novel genes into the genome. Within the biotechnology industry, there is increasing interest in the exploitation of algal species as sources of valuable biochemicals and pharmaceuticals and in processes such as fuel and polysaccharide production, bioremediation, and aquaculture. These industrial applications would benefit tremendously from genetic engineering technology that allows the manipulation of biochemical pathways leading to higher product yields and improved strains. The development of DNA transformation for algal species is still in its infancy.

Genetic engineering of algae is not common due to problems related to the design of vectors (i.e. plasmids or viruses) that can be successfully incorporated into the algae, accepted by the cell and expressed in a satisfying way. Most studies have therefore been performed on the green alga *Chlamydomonas reinhardtiiand.* (Mostafa and Sheekh, 2005). In particular, the development of nuclear and organellar transformation systems for the green alga *Chlamydomonas reinhardtiihas* led to a wealth of studies on gene function and expression in this organism. Success with *Chlamydomonas* has also acted as a catalyst in the development of transformation systems for other algae, most notably the related green alga *Volvox carteri* and several diatom species. One of the major problems when applying genetic engineering on new kinds of organisms is that of designing specific vectors that can be transformed into the cells, accepted by the cell and expressed in an adequate way. (Cerutti et al., 1997). The ability to introduce and achieve desired levels of expression of foreign genes has been made possible by technical development for the incorporation of DNA into algal cells. Techniques used in the transformation of algal cells include injection of DNA through fine glass needles (Able et al., 1989) (microinjection), bombardment of cells with DNA-coated gold particles, and virus infection. Other methods used to make the cells prepared for uptake of DNA fragments or plasmids are the use of an electrical charge to temporarily open pores in the cell membrane (electroporation), or agitation of algal protoplasts, i.e. algae without cell walls, with glass beads. (Kindle et al., 1990). Promoter systems have been developed so that the algal cells can express the introduced DNA in a satisfying way. Homologous promoters are usually preferred, since heterologous promoters (those from other organisms) sometimes do not drive the expression of the transformed genes in an efficient way. Another important factor is the selection of reporter genes, which identify the cell that has been successfully transformed. In bacteria, genes conferring antibiotic resistance are the most widely used reporter genes. Usually antibiotic resistance genes are not used as reporter genes in algae due to their often-natural resistance to antibiotic compounds. Reporter genes that have been used include the gene that encodes the enzyme arylsulphatase. This enzyme is normally expressed under sulphur starvation and it causes the algal cells to produce an easily detectable coloured substance. Pesticide resistance is popular as a selective marker in plant genetic engineering, and can probably also be used as such in similar studies on algae.

5.2 STRATEGIES FOR METABOLIC ENGINEERING IN ALGAE

The introduction of exogenous DNA into an algal cell and the stable expression of that DNA involve numerous steps, each of which must be achieved for the species under study. First, the organism must be available as an axenic, clonal culture, and conditions for its maintenance and growth in the laboratory must be established. Second, techniques must be developed for the regeneration of the organism from single cell. This makes possible the selection of clones arising from the transformation of a few individual cells within a population. For unicellular algae this is not a problem, but for filamentous or multicellular algae, cell and tissue techniques must be developed that allows their growth from cells or calli on nutrient agar plates. An alternative approach would be the generation of the organism from transformed gametes. The third consideration is the method by which cloned DNA is delivered into the cell. The following DNA transfer techniques have proven successful in algal transformation:

(i) The biolistic process, which involves high-velocity bombardment of cells or tissue with DNA-coated microparticles,
(ii) Microinjection,
(iii) Electroporation of cells or protoplast suspensions,
(iv) Agitation of cell wall-deficient cells with DNA and glass beads, and
(v) Agitation of walled cells with DNA and silicon carbide whiskers.

Researchers have focused on various techniques for the genetic manipulation of several microalgal species. Some case studies are presented here.

Chlorella

The unicellular green alga *Chlorella* has long been a favorite experimental organism for studying the physiology, biochemistry, metabolic activity, biomass production and molecular biology of a number of cellular processes. A little has been done on *Chlorella* transformation by foreign genes (Nagels *et al.*, 1989). *Chlorella* belongs to eukaryotic organisms that undergo a simple cell division cycle, in most cases without a sexual-type stage, enabling them to complete their cell cycle within a few hours and making genetic selection and strain screening relatively quick and easy. This also allows much more rapid development and demonstration of production processes than with other agriculture crops. The *Chlorella* nuclear genome has a number of characteristics in common with the nuclear genome of certain monocots (e.g., maize), such as (i) highly methylated DNA and high GC content, (ii) genes containing many introns and higher-plant consensus intron/exon splice sites, and (iii) codon usage bias with a preference for C in the third position of the triplet codons. Therefore, *Chlorella sorokiniana* with its doubling time at 39°C may prove to be a useful eukaryotic micro-organism in which to express higher plant genes/cDNAs for the production of compounds of medical or industrial importance (Mostafa and Sheekh, 2005). Some literature reports of genetic engineering of Chlorella are given in Table 5.1.

Table 5.1 Literature reports on *Chlorella*

Microalga	Major Research Breakthrough	Reference
Chlorella ellipsoida	Introduced a luciferase gene construct into protoplasts by incubation with polyethylene glycol and were able to detect luciferase activity in the cells.	Jarvis and Brown, 1991
Chlorella sorokiniana	Used electroporation and proved that *C .sorokiniana* has a NADP-specific glutamate dehydrogenase (NADP-GDH) nuclear-gene encoding a precursor mRNA.	Miller et al., 1994
Chlorella vulgaris	Established the conditions required for the transformation of *Chlorella vulgaris* by electroporation	Chow and Tung, 1999

Volvox

The green alga *Volvox* represents the simplest multicellular organism. *Volvox* is composed of only 2 cell types, somatic and reproductive. *Volvox* species represent a simple and attractive model system for studying the transition from unicellular eukaryotes to multicellular organisms. Nitrate assimilation is the major route by which inorganic nitrogen is converted, under normal conditions, to the end product ammonia and incorporated into amino acids. Assimilatory nitrate reductase catalysis is the first and rate limiting step in the pathway which converts nitrate to nitrite, followed by a reduction of nitrite to ammonia catalysed by nitrite reductase (Mostafa and Sheekh, 2005). Thus, *V. carteri* becomes one of the simplest multicellular organism that is accessible to detailed molecular studies of genes regulating cellular differentiation and morphogenesis (Table 5.2).

Table 5.2 Literature reports on *Volvox*

Microalga	Major Research Breakthrough	Reference
Volvox	Developed 2 reporter gene systems for *Volvox* based on the genes for arylsulphatase (ARS) and sexual pheromone.	Halmann and Sumper (1994)
Volvox carteri	Transformation of *V. carteri* using the nitrate reductase gene. Very high expression obtained due to identification of introns of the gene.	Gruber et al. (1996)
Volvox	Demonstrated the expression of a foreign gene from Chlorella in *Volvox* and development of an important new marker for transformation.	Hallman and Sumper (1996)

Diatoms

Diatoms are a very large and diverse group of microalgae found in the sea, fresh water and terrestrial environments. Not only are they of tremendous importance as primary producers responsible for a significant proportion of worldwide oxygen production and carbon dioxide fixation in aquatic environments, but they are also of commercial value as sources of fuels and specialty oils, biopharmaceuticals, and aquaculture feedstock (Lebeau and Robert, 2003). Recent work has led to the development of reliable nuclear transformation procedures for several species, thereby expanding the usefulness of diatoms as experimental systems and as a commercial resource. Some genetic modifications have been presented in Table 5.3.

Table 5.3 Literature reports on diatoms

Microalga	Major Research Breakthrough	Reference
Cyclotella cryptica	Production of lipid-derived biodiesel fuel from microscopic algae by overexpression of ACCase	Rossler et al., (1994)
Phaeodactylum tricornutum	Phaeodactylum tricornutum was genetically engineered to thrive on exogenous glucose in the absence of light through the introduction of a gene encoding a glucose transporter (glut 1 or hup 1).	Rossler et al., (1994)
Cyclotella cryptica and *Navicula saprophila*	Successfully transformed the 2 species, using *E. coli* gene that encodes the enzyme acetyl-CoA carboxylase. This DNA construct was introduced into the algal cells by microprojectile bombardment.	Dunahay et al., (1996)
Cycloteus cryptica	Modified the lipid metabolism by introducing additional copies of the gene for acetyl-CoA carboxylase, thereby increasing the activity of this key regulatory enzyme.	Dunahay et al., (1996)
Navicula pelliculosa	Described a transformation based approach to investigate intracellular targeting and function of frustulins found in the cell walls.	Fischer et al., (1999)
Phaeodactylem tricornutum	Developed a transformation system for *P. tricornutum* based on microparticle bombardment.	Zaslavskaia et al., (2000).

The successful transformation of marine diatoms can be attributed to several factors.

- The first factor was the use of microparticle bombardment as a DNA delivery system able to penetrate the diatoms siliceous cell walls. Earlier attempts by Dunahay et al., (1995) to introduce DNA into *Cyclotella cryptica* by electroporation were met with limited success.

- Second, careful choice of markers has allowed transformant selection in organisms that are diploid and are naturally resistant to a wide range of antibiotics and herbicides (Apt et al., 1996).
- Finally, diatom promoters were used to drive expression of the introduced genes since heterologous promoters work poorly, if at all, in the microalgae (Dunahay et al., 1995).

Other Algae

The work of Neuhaus et al. (1986) has demonstrated the feasibility of transforming the giant unicellular, uninuclear green alga *Acetabularia acetabulum* (mediterranea) by microinjection of isolated nuclei and their subsequent implantation into anucleate cell fragments. Geil and Mandoli (1996) have evidence for transformation of *A. acetabulum* gametangia by microparticle bombardment with gene constructs for β-glucuronidase or neomycin phosphotransferase. Porath et al., (1997) reported evidence of both transient and stable expression of transgenes in *Dunaliella salina.* This preliminary evidence for stable transformation suggests that Dunaliella, like the marine diatoms, will soon be subject to molecular analysis of the biochemical pathways that are of biological interest and commercial value. Luhring and Witzemann (1995) injected RNA molecules encoding rat muscle proteins into the giant internodal cells of *Chara corallina.* The first report of transformation of a macroalga was from Kurtzman and Cheney (1991), who demonstrated the transient expression of the *E. coli* β-glucuronidase gene (gus) in the carrageenan-producing rhodophyte *Kappaphycus alvarezii.* Cheney et al., (2001) has reported the stable genetic transformation of the multicellular, macroscopic marine red alga (or seaweed) *Porphyra yezoensis.*

5.3 APPLICATIONS

Genetic engineering has proved to be an important tool in improving the production of various metabolites in microalgae. It has also contributed in the genetic breeding of algae to obtain highly productive strains. Since Algae are a source of many important compounds as discussed in the earlier chapters, a lot of research has been focused on genetic modifications for higher yield of the target metabolite. Some of these attempts by various researchers have been presented in Table 5.4.

Another area where genetic engineering has helped is in the cultivation of marine macroalgae. The brown alga *Laminaria japonica* was originally introduced to China from Japan and is cultured there extensively. In China there is no natural population of this kelp due to high summer temperatures. Genetic breeding of *L. japonica* has been conducted since the 1950s and has created a few highly productive strains. This has increased the annual production from 10 t dry weight in 1952 to the current 350,000 t. *L. japonica* is

Table 5.4 Genetic modifications for improving metabolite yield

Metabolite	Major Research Breakthrough	Reference
Astaxanthin	The gene coding for the enzyme that converts β-carotene into astaxanthin was identified from *H. pluvialis* and cloned into *Synechococcus*.	Harker and Hirschberg, 1997.
Lycopene (carotenoid pigment) and ubiquinone (major component of the aerobic respiratory chain)	The dxps gene from the cyanobacterium *Synechocystis sp.* was sequenced and cloned into a plasmid. Cells of *E. coli* were transformed with the plasmid and the dpxs. gene was expressed.	Harker and Bramley,1999
Astaxanthin	The gene coding for the enzyme that converts β-carotene into astaxanthin from *H. pluvialis* was cloned in *E. coli* and the expression was higher than for the untransformed cells.	Wang et al., 1999.

consumed as a subsidiary food and used in the production of iodine mannitol and alginate. Genetic engineering is expected to be an effective means to develop kelp as a marine bioreactor to produce oral drugs such as vaccines. (Qin et al., 1999) Still successful transformation in kelp has been reported but models for the insertion of foreign genes have been set up. The model includes introduction of foreign DNA by biolistic bombardment, i.e. bombarding cells with microprojectiles coated with DNA and the use of the SV40 promoter as a transcription initiator to drive the gene expression.

Chapter **6**

MATHEMATICAL MODELLING

6.1 INTRODUCTION

Microalgal cultivation for the production of high value metabolites requires economical and large scale culturing of microalgal biomass. In the previous chapters, different types of cultivation techniques have been discussed in detail.There are number of advantages of closed over the open systems. Hence, much attention is being given to photobioreactors. Consequently, many models have been developed for them. Modeling involves a systematic mathematical analysis of the system. Considering the extremely high cost of production, microbial process engineering approach to experimental work can make an efficient use of lab scale experimentation employing a scaled down model system. Modeling is the process in which the analyst constructs a set of mathematical relationships together with the boundary conditions and initial conditions that are isomorphic to relationships among the process variables. Because of the complexity of the real system (physico-chemical, physiological, biochemical and genetic laws) and the mathematical relations, whatever model is developed, is bound to be highly idealized and generally gives a faithful representation of only a few process properties. The first equation of the model is most often a simple version of the mass conservation law. The models can be generally categorized as follows.

- **Structured and Unstructured Models**

  In an unstructured model, the reaction rate is described by input and output rates that are constants or functions of environment. These models don't include variations of intracellular properties. If the model includes the functions of intracellular structures, it is considered as a structured model.

- **Segregated and Non-segregated Models**

  If all cells are considered to have equal properties the model is said to be non-segregated. Introduction of age distribution, size distribution, morphology etc., makes it a segregated model.

The formulation of mathematical biological process models, from the standpoint of systems analysis is usually realized in three stages.

(i) Quantitative analysis of the system structure, usually based on the knowledge of metabolic pathways and biogenesis of the desired product.

(ii) Formulation of the model in a general mathematical form. This stage is sometimes called structure synthesis of process functional operator.

(iii) Identification and determination of numerical values of model constants and/or parameters, which is based upon experimental or other operating data from a real process. The experimental data obtained, is fitted to the models. Many computer programs are available for fitting experimental data, solving ordinary differential equations, partial differential equations and regression analysis (Volesky and Votruba, 1992).

For microalgae, recent biotechnological and technical advances necessitate a reappraisal of their possible future contribution in areas of food, cosmetics, pharmaceuticals, feed, agriculture, aquaculture and the environment. The design of technical and technological basis for photobioreactors is the most important issue for economic success in the field of phototropic biotechnology. For future applications, open pond systems for large-scale production have lower potential than the closed systems. For high value products, which usually are metabolites, closed system of photobioreactor seem to be the promising field for technical developments. Hence, much emphasis is being given to PBRs for high value metabolite production. The use of activities of photosynthetic microorganisms, calls for quantitative understanding of their behavior and kinetics. This is often the rate-controlling process. The cultivation of photosynthetic microorganisms, unlike others, needs specific problems to be treated, leading to a particular model. These models can be applied to process development, process optimization or process control. The state of art in modeling photosynthetic culture of microorganisms includes widely made assumption that growth rate is closely related to light energy available inside the medium. The light energy transfer through a dense medium within a reactor is generally uneven. Therefore, this requires special attention. Other problems like CO_2 transfer, nutrient transfer and hydrodynamic stress also need to be considered.

6.2 MODELS FOR MICROALGAL CULTIVATION

Many models have been proposed for optimization of microalgal growth, addressing the problems discussed. Some of them are discussed here.

Light Transfer

Mathematical modeling of photobioreactors requires coupling between the metabolism and the physical phenomena of light transfer inside the culture medium. The study of interaction between physical limitation by light and biological limitations in photobioreactors leads to very complex partial differential equations. Schuster proposed a simplified, monodimensional equation of radiative transfer, incorporating two independent parameters related to light absorption and scattering. Analytical solutions of these equations offer solutions for the determination of local radiant energy available in the culture medium and approximate quantification of the total energy absorbed in the vessel. The following assumptions were made

- Liquid medium taken is non-emitting and non-fluorescing.
- The light dissipation depends upon two independent phenomena, namely; absorption by pigments and scattering by the whole cells (called shading effect).
- Light energy flux in monodimensional geometry.

The specific light intensity of wavelength l, is given by $I_l(z, \theta, \phi)$. At each point of z-axis, the direction making an angle θ with the z-axis, with φ as the azimuthal angle in spherical coordinates, characterizes the radiation field. Assuming the radiation field as isotropic, the mean intensity $J_{z\lambda}$ is calculated by

$$J_{z\lambda} = \frac{1}{4\pi}\int_0^{2\pi}\int_0^{2\pi} I_\lambda(z,\theta)\sin\theta.d\theta.d\phi. \qquad (6.1)$$

The radiant energy flux in z-direction similarly is calculated by

$$E_\lambda = \frac{1}{4\pi}\int_0^{2\pi}\int_0^{2\pi} I_\lambda(z,\theta)\sin\theta.\cos d\theta.\ d\phi. \qquad (6.2)$$

These parameters are necessary to characterize the path of the light as medium is assumed to be absorbing and scattering. The assumption that absorption and scattering are proportional to the biomass concentration introduces the absorption and scattering mass coefficients, $E_{a\lambda}$ and $E_{s\lambda}$.

The variation of the radiant energy flux along the z-axis is related to the mean intensity,

$$\frac{dF_{z\lambda}}{dz} = -\,4\pi\, E_{a\lambda}\, C_x J_{z\lambda} \qquad (6.3)$$

The right hand term represents the local absorption rate noted, A_{zl}, so that,

$$A_{zl} = 4\pi\, E_{a\lambda}\, C_x J_{z\lambda} \qquad (6.4)$$

Radiant energy on the elemental volume of thickness Äz and their integration gives the following equations. The radiant energy flux F_z is given by the vectorial sum of the direct and reverse fluxes, so that,

$$F_z = F_z^+ - F_z^- \qquad (6.5)$$

J_z is proportional to sum of F_z^+ and F_z^-,

$$J_z = \frac{1}{4\pi}(F_z^+ - F_z^-) \tag{6.6}$$

Substituting the values for F_z^+ and F_z^-, we get

$$\frac{4\pi z}{F_0} = 2.\frac{(1+\alpha)e^{-\delta(z-1)} - (1-\alpha)^{\delta(z-1)}}{(1+\alpha)^2 e^{\delta} - (1-\alpha)^2 e^{-\delta}} \tag{6.7}$$

where, $\alpha = [E_a/(E_a + E_s)]^{1/2}$

$\delta = (E_a + E_s).\ a.\ C_x.L$

and z is the dimensionless abscissa z/L.

Growth kinetics and light transfer models can be coupled as follows. Growth rate versus radiant energy flux F_z generally follows a Monod-type law. Considering the total available energy $4\pi J_z$ growth rate can be given by

$$r_x = \mu_m.\ C_x.\ \frac{4\pi J_z}{K_j + 4\pi J_z} = \mu C_x. \tag{6.8}$$

For low light energy inputs, incident light energy is negligible compared to K_j, which means that specific growth rate is proportional to $4\pi J_z$. Then,

$$r_x = \mu_m.C_x.\ \frac{4\pi J_z}{K} \tag{6.9}$$

the mean volumetric growth rate can be adequately calculated assuming Schuster's approximations. The limiting cases can be examined as follows.

If the biomass concentration is low: $r_x = \mu_m.C_x.F/K_j$ (6.10)

Although the rate of metabolism is limited by light energy, growth is exponential with an apparent maximum growth rate π_m. The ratio F/K_j is proportional to incident energy and independent of the absorption and scattering characteristic of the medium.

If the biomass concentration high: $r_x = (\mu_m/K_j)(F_o/E_a.L)(2a/1+\alpha)$ (6.11)

The growth rate is independent of the biomass concentration, indicating a linear growth, the rate of which is strongly dependent on absorption and scattering characteristics of the medium (E_a and a) and proportional to the incident energy F_o. The exponential growth (r_x/μ_m proportional to C_x) and linear growth (r_x/m_m independent of C_x) are conserved, whatever the ratio of F_o/K_j.

Temperature, Light and Growth

It is difficult to study potential microalgal candidates under all conditions of light and temperature. Mathematical modeling of these variables provides a useful tool for growth prediction and characterization. Photic and thermal optima and tolerances of photosynthetic aquatic microorganisms are obviously dependent on their physiological capabilities, which are species specific and depend upon environmental factors.

Various authors have sought mathematical model for microalgal growth but light and temperature affects are usually considered independently. It is therefore better to study the simultaneous effect of light and temperature on growth (Cornet at al., 1992). It is assumed that cultures are grown in media providing non-limiting nutrients. In practical terms, the modeling equation must be able to satisfactorily represent the observed photoinhibition behavior of the three species at the various temperatures studied. The function must minimize the difference between observed and predicted values in order to match as closely as possible general shape of experimental results. Additionally, the function must rely upon a restricted number of coefficients that can be expressed as experimentally measurable entities (optimal light energy, maximal growth rate, etc.). The use of combination of straight line segment is also eliminated. Since the derivative is undetermined at their intersection, one would face difficulties later when using such a function in a mathematical model. Therefore Peeters' formula (Eilers and Peeters, 1988) is chosen to represent the growth of microalgae as a function of light energy.

$$\mu_T = 2.\ \mu_{Tmax}.(1+\beta).\frac{x}{x^2+2\beta x+1} \tag{6.12}$$

in which, $x = I/I_{opt}$

In this equation I is the light energy, I_{opt} is the optimal light energy for growth and β is the sharpness coefficient.

The curves obtained in this equation represent response to the light intensity and temperature variations. It is therefore essential to express growth by a function $\mu = f\,(I, T)$, which represents the simultaneous effect of the two factors. We know that μ_{max} and I_{opt} depend upon the temperature. We can similarly consider that the same holds true for β. In order to represent the variations of three coefficients, equation can be written as follows.

$$y = y_{max}.\left(\frac{T}{\theta_{opt}}.\ w^u\right)^{\alpha} \tag{6.13}$$

in which,

$$w = \frac{\theta_0 - T}{\theta_0 - \theta_{opt}} \tag{6.14}$$

and

$$u = \frac{\theta_0 - \theta_{opt}}{\theta_{opt}} \tag{6.15}$$

where T is the temperature, θ_{opt} the temperature at which m_{max}, I_{opt} and β are maximum, θ_o the temperature at which m_{max}, I_{opt} and β become zero ($\theta_o > \theta_{opt}$) and a is a power.

Fitting is done by a gradient method, applying the method of least square. The fluctuations in μ_{max} and corresponding I_{opt} with temperature, illustrate the interdependence between light and temperature and necessity of simultaneous expression of these two.

CO_2 Concentrating Mechanism

Detailed structural mathematical models have been developed for the functioning of CO_2 concentration mechanisms (CCM), taking into account cell and chloroplast structure (Fridlyand L., 1997; Giersch. C., 2003). Many microalgae, when growing under low dissolved inorganic carbon concentration (C_i), acquire the capacity of efficient utilization of CO_2. This is caused by a specific CO_2 concentrating mechanism. CO_2 is the only form of C_i fixed by ribulose-1,5-biphosphate carboxylase/oxygenase (rubisco). CO_2 is increased at the location of rubisco during CCM. However, the cells have to overcome some difficulties. Firstly, bicarbonate is the only possible source of CO_2 evolution inside the chloroplast if C_4 mechanism is absent. However, the plastid stroma is alkaline in the light and thus the equilibrium of introconversion of CO_2 and bicarbonate is directed mainly from HCO_3^-(HC). Therefore, a sufficient rate of CO_2 evolution should be provided in unfavorable alkaline medium. Secondly, CO_2 diffusion away from the site of its concentration diminishes the efficiency of CCM. Thus, decrease in outward flow of CO_2 is the evolutionary goal of cell. The difficulties in limiting CO_2 diffusion are associated with rather easy penetration of CO_2 through membrane. It is impossible to ensure a low value of CO_2 permeability only by chloroplast envelope resistance that does not exceed the diffusive resistance of the two-lipid bilayer membrane. The induction of CCM in microalgae is accompanied by the increased carboxylase anhydrase (CA) activity in a number of cell compartments. Only the considerable CO_2 permeability coefficient for a double lipid layer membrane obtained is used for plasma lamella and every other membrane. The following assumptions are made

- Structural model of microalgae is assumed as pyrenoid, chloroplast stroma, cytoplasm and periplasmic space.
- Cell is considered to be spherical for the purpose of modelling.
- Chloroplast volume around the pyrenoid is considered as sphere that has several layers with different CO_2 permeability coefficient.
- The chloroplast volume is assumed to be 40% of the cell volume.

The transformation rate (V_p) of bicarbonate into CO_2 inside thylakoid is,

$$V_p = E_H [HC] \tag{6.16}$$

The pumping mechanism that transports bicarbonate actively into a chloroplast through a membrane may be expressed as,

$$V_a = \frac{V_t[HC]}{[HC]+K_a} \tag{6.17}$$

CO_2 assimilation inside a pyrenoid (V_s) is given by

$$V_s = \frac{V_t[CO_2]}{[CO_2] + K_c(1 + [O_2]/K_0)} \tag{6.18}$$

CA accelerates the reaction of interconversion between various C_i forms (the main reaction is CO_2! $H^+ + HCO_3^-$ at pH. 5.0–8.0. Its action is given by the equations;

$$V_b = \frac{V_{ba}K_{ca}[HC]}{K_{ba}K_{ca} + K_{ca}[HC] + K_{ba}[CO_2]} \tag{6.19}$$

$$V_c = \frac{V_{ca}K_{ba}[HC]}{K_{ba}K_{ca} + K_{ca}[HC] + K_{ba}[CO_2]} \tag{6.20}$$

where V_b and V_c are the rates of bicarbonate dehydration and CO_2 hydration respectively.

CO_2 and HCO_3 diffusion fluxes per cell (J_c and J_b) can be given according to Fick's law,

$$J_b = P_c([CO_2]_i - [CO_2]_k) \tag{6.21}$$

$$J_c = P_c([HC]_i - [HC]_k) \tag{6.22}$$

where k and i designate the inner and outer borders of certain layer, P_c and P_h are CO_2 and bicarbonate permeability coefficients through a layer of a cell.

For i the spherical layer the CO_2 permeability coefficient (P_i) can be written as,

$$P_i = \frac{4\pi D_1}{1/R_i - 1/R_{i+1}} \tag{6.23}$$

Where D_i is the diffusion coefficient for the i th layer, R_i and R_{i+1} are the inner and outer radii of the layer. If there are n layers, generally permeability coefficient is given by,

$$P = (1/P_1 + 1/P_2 + 1/P_3 + \ldots + 1/P_n)^{-1} \tag{6.24}$$

The mean CO_2 diffusion coefficient through overall thylakoid layers can be determined as,

$$D_{ac} = (1/D_{sc} + N/P_{mc})^{-1} \tag{6.25}$$

The simplest possible CCM incorporates CO_2 diffusion through the unstirred layer and the plasma lamella and CO_2 hydration in cytoplast, in presence of CA. Then, bicarbonate should be actively transported into a chloroplast, where CO_2 evolution takes place in pyrenoid. It is supposed that CA is absent on the cell surface. The system of differential equations for the changes in the concentration of C_i species and O_2 in various compartments can be written as

$$d[CO_2]_p/dt = (V3 - V5cp - V4)/T_1$$

$$d[HC]_p/dt = (V7up - V3)/T_1$$

$$d[O_2]_u/dt = (V8 - V9)/T_2$$

$$d[HC]_u/dt = (V6cx - V7up)/T_2$$
$$d[CO_2]_c/dt = (V5cp - V5mc - V1c - V2c)/T_3$$
$$d[HC]_c/dt = (V2c - V1c - V6cx)/T_3 \qquad (6.26)$$

where subscripts p, u and c designate pyrenoid, the stroma near the inner border of chloroplast and the cytoplasm.

Inorganic Carbon Release during Photosynthesis

A model of inorganic carbon release by unicellular algae is proposed based on the involvement of HCO_3/NO_3 antiport growth under non-steady state conditions, such as nutrient pulsing (Collos, 1997).

From a cultural work on unicellular algae, it has been shown that algal dark respiration varies from 5 to 60% of gross photosynthesis. Respiration in the light can be very different from that in the dark. For example, $^{14}CO_2$ efflux from cells of unicellular algae can be five times higher in the light than in the dark. The model has been proposed on the available physiological knowledge of the relationships between carbon and nitrogen metabolism in plant cells. The present model can provide an indirect way to estimate rates of inorganic carbon loss during photosynthesis in phytoplankton. The following assumptions are made:

- An exchange of nitrate ion for every bicarbonate ion excreted i.e., an overall balance for electroneutrality of the medium.
- This overall balance does not discriminate between dark respiration photorespiration, because it refers to specific mechanism (ion exchange).

The model describes respiration as follows:

$$R = r_1 P_{max} + r_2 P_g \qquad (6.27)$$

The respiration rate is proportional to the maximum photosynthetic rate (P_{max}) and to gross photosynthesis (P_g) with a basic respiration constant (r_1) representing 'maintenance' respiration and a second constant (r_2) representing 'growth' respiration or the cost of synthesis. The model postulates an exchange of nitrate ion for every bicarbonate ion excreted, the latter coming from decarboxylation of malate formed during nitrate reduction. Such a model has been challenged on a number of grounds, but should apply best to unicellular algae where sites of nitrate uptake and malate production are close to each other.

If V_{Co} and V_{Cn} represent inorganic carbon uptake in absence and in presence of nitrate respectively (both expressed on a molar basis), then the following equivalence can be written:

$$V_{Co} - V_{Cn} = kV_{NO_3} \qquad (6.28)$$

where V_{NO_3} is the nitrate uptake rate and k is proportionality constant which should be equal to one, if inorganic carbon release is due only to nitrate uptake.

The respiration coefficient r_2 in Eq. (4.1) representing inorganic carbon release during growth on nitrate can be expressed as,

$$r_2 = (V_{Co} - V_{Cn})/V_{Co} \tag{6.29}$$

Phosphate Transfer and Assimilation

The operation of a three-compartment phosphate model of microalgal growth (phosphate interaction model, PIM), containing internal P pools for polyphosphate (PolyP), soluble inorganic P (SIP) and structural and other organic P (SOP), is compared with that of a conventional quota model, which contains only a single internal nutrient pool (John and Flynn, 2000). Phosphorous (P) is an important rate-limiting nutrient in many biosystems. The development of multi-nutrient (e.g., N – P) models for phytoplankton could have both pure and applied applications. However, an important question concerns the level of model complexity required to achieve an adequate simulation. There is, however, a major problem in developing mechanistic P models, namely parameterization of the internal pools of phosphate and the flows between them. The reason is the multiplicity of roles for P and the complexity of the fractionation process required to track them. A three compartmental model is assumed containing phosphorous (P) as external P, PlolyP, SIP and SOP. In addition to an internal pool of orthophosphate, it is divided into the intracellular P fractions, according to their functions into structural (DNA and membrane lipid-P), synthetic (acid-soluble polyphosphate intermediates, nucleotides, and acid insoluble RNA-P) and finally storage pools (comprising mostly of acid-soluble and acid-insoluble polyphosphates). Inorganic polyphosphate (PolyP) may be accumulated to high concentrations a phenomenon termed 'polyphosphate overplus'.

Nutrient Limited Growth

Many models of microalgal growth focus on steady state carbon budgets. This approach can be very useful if one's primary concern is to predict growth on the basis of a limited set of measurements. However, it is less likely to be successful if one purports to understand the regulation growth in response to variations in the nutrient concentrations in the environment. A generic feature of many microalgal growth models is their focus on carbon budgets. For instance,

Net increase = (Photosynthesis – Respiration) – Loss terms

Since both light and nutrients may be growth limiting, it seems logical to give them equal status. However, to date no single model does this. While models on light dependent growth may be fairly elaborate, models on limited growth are often relatively simple. Zonneveld, 1996 have proposed a model for nutrient limited growth. They report a model for microalgal growth on two, potentially growth limiting nutrients. Two assumptions namely, nutrient limited growth and a well-known Droop's equation are made. Droop's equation describes the specific growth rate as dependent on the cell quota (Q) of a limiting nutrient and can be derived from simple rules on uptake and utilization of the particular nutrient. The total amount of nutrient of a cell, Droop's cell quota Q, can be subdivided into two parts. Some nutrient is incorporated in the structural biomass; this corresponds to Droop's subsistence quota k_Q, the minimum quota necessary for life. When cells are grown at saturating concentrations of nutrient in the environment for sufficiently long time, the nutrient density of the cells eventually reaches a constant value $[N_l^m]$, which cannot be exceeded. Intracellular density will be expressed as a fraction of this maximum; $n_l = [N_l]/[N_l^m]$. This dimensionless quantity takes values between 0 and 1. Nutrient uptake rate (1) is proportional to cell volume, (2) varies inversely with intracellular nutrient density and (3) depends hyperbolically on the environmental concentration x_i (i.e. Michaelis-Menten uptake kinetics).

$$A_l = \left[A^l{}_m\right].\frac{V.x_l}{n_l(k_l + x_l)} \tag{6.30}$$

In this equation x_l denotes the concentration of the limiting nutrient in the environment and k_l is the half saturation concentration. The parameter $[A_l]$ denotes a nutrient uptake rate per unit volume under saturating external nutrient concentration, if the intracellular nutrient density is at its maximum. Assume that the rate of change of nutrient density can be described by;

$$\frac{d[N_L]}{dt} = \frac{A_l}{V} - v_l[N_l] \tag{6.31}$$

The quantity $v_l[N_l]$ represents usage of nutrient in the transient pool; v_l is a proportionality constant. The utilization rate equals the assimilation rate A (given by Eq. 6.1) minus the rate of change of the amount of nutrient, $N_l = [N_l]V$. On the other hand, if nutrients are only used for broth (no leakage, no destruction) and nutrient costs per unit volume $[G_l]$ are constant, one knows that the utilization rate must equal $[G_l](d/dt)V$, so that, equation can be solved for the growth rate $(d/dt)V$. From the above equation, this is found to be, (6.32)

$$\mu \equiv \frac{1}{V}.\frac{dV}{dt} = v_l \frac{[N_l]}{[G_l]+[N_l]} \tag{6.33}$$

For two or more nutrients, the one with the smallest ratio determines the growth rate. The utilization rate is given by $[G_i]\ \mu_i V$ (i ranges over the potentially limiting nutrients). m's for the different potentially limiting nutrients, i.e. ;

$$\mu = \min\{\mu_i\} \text{ with } m_{l=} v_l \frac{[N_i]}{[G_i]+[N_i]} \tag{6.34}$$

Concerning the uptake of non-limiting nutrient it can be assumed that (1) uptake rate is proportional to volume, (2) depends hyperbolically on nutrient concentration in the environment, and (3) is proportional to density $[N_l]$ of the limiting nutrient.

Uptakes of non-limiting nutrients will depend on the limiting nutrient. Density of non-limiting nutrient;

$$\frac{d}{dt}[\mathrm{N}_n] = \left[A_n{}^m \frac{[N_l]}{[N_l{}^m]} \right] f_n - \mu([G_n]+[N_n]) \tag{6.35}$$

The parameter $[A_n{}^m]$ represents the uptake rate per minute volume, at saturating concentration of the non-limiting nutrient, with maximum nutrient density of the limiting nutrient. Four experiments were devised in such a way that phosphorous was limiting in two of them, and vitamin was limiting in the other two and the experimental data was found to fit the model very well.

6.3 BIOSORPTION EQUILIBRIUM AND KINETIC MODELS – CASE STUDY

The biosorption process involves a solid phase (sorbent) and a liquid phase (solvent, normally water) containing a dissolved species to be sorbed (sorbate, e.g. metal ions). Owing to the higher affinity of the sorbent for the sorbate species, the latter is attracted to the solid and bound by a number of different mechanisms. This process continues until equilibrium is established between dissolved and solid-bound sorbate (at a residual, final or equilibrium concentration C_{eq}). The affinity of the sorbent for the sorbate determines its distribution between the solid and liquid phases. The quality of the sorbent material is judged by how much sorbate it can attract and retain in an immobilized form. The determination of the metal uptake (q) by the biosorbent is most often based on the material balance of the sorption system: sorbate that disappeared from the solution must be in the solid. The sorption uptake, q, can be expressed in different units depending on the purpose of the exercise: for example, milligrams of metal sorbed per gram of the (dry) sorbent material (the basis for engineering process-mass balance calculations), or m mol g^{-1} or meq g^{-1} (when stoichiometry and/or mechanism are considered). For (bio)sorption process scale-up and applications, the uptake expressed per unit (reactor) volume is also important.

Which sorbent is 'better'? There is no direct answer to this question until it is qualified: at what residual concentration? The uptake of metal by the two biosorbents must be compared only at the same equilibrium concentration. This is the reason why one comparison at low C_{eq} (e.g. 10 mg l^{-1}) and another at high C_{eq} (e.g. 200 mg l^{-1}) are made in some biosorption screens (Holan et al., 1993; Holan and Volesky, 1994). The comparison of single-sorbate sorption performance is best based on a complete single-sorbate sorption isotherm curve derived under the same environmental conditions (e.g. pH, temperature, ionic strength). Sorption isotherms are plots of the sorption uptake (q) and the final equilibrium concentration of the residual sorbate remaining in the solution (C_{eq}). The classical models of Langmuir (Langmuir, 1918) (Equation 6.36) and Freundlich (Freundlich, 1907) (Equation 6.37) are often used to describe the relationship.

$$q_{eq} = q_{eq}\,\frac{K_L.C_{eq}}{1+K_L C_{eq}} \tag{6.36}$$

$$q_{eq} = K_F.C_{eq}^{\ n} \tag{6.37}$$

These models, based on the original early work on adsorption to activated carbon, do not reflect any mechanisms of sorbate uptake; the equations are just capable of reflecting the experimental curves. Ion exchange processes apparently play an important role in biosorption, and this is reflected in correspondingly different equilibrium models proposed for biosorption based on ion-exchange principles (Schiewer and Volesky, 1997). However, although the Langmuir isotherm model does not correspond to the biosorption (ion-exchange) phenomena, it is being widely used. It incorporates two easily interpretable constants: q_{max}, which corresponds to the highest possible sorbate uptake (the complete saturation isotherm-curve plateau); and coefficient K_L, which is related to the affinity between the sorbent and the sorbate. Low values of K_L are reflected in the steep initial slope of a sorption isotherm, indicating a desirable high affinity. Thus, for 'good' sorbents in general, one is looking for a high q_{max} and a steep initial sorption isotherm slope (i.e. low K_L). It is relatively simple to obtain laboratory equilibrium- sorption data for a single sorbate (a small amount of the sorbent to be tested is brought into contact with a solution containing the given sorbate). However, the environmental parameters in the sorption system (particularly pH) may have to be controlled carefully unless an appropriate (ion-exchange) sorption model is available to compensate for the changing proton concentration in the system (Schiewer and Volesky, 1997). Otherwise, the pH must be controlled over the entire period of contact until the sorption equilibrium is reached. This may take a few hours or much longer, depending on the size of sorption particles and the time it takes until they attain sorption equilibrium. A simple preliminary sorption kinetics test will establish the exposure time necessary for the given sorbent particles to reach the equilibrium state (characterized by unchanging sorbate concentration in the solution) by using time-based analyses. Enough time must be allowed for the sorption system to reach equilibrium. Following the sorption contact, q could also be determined directly by analysing the separated solids and thus closing the material balance on the sorbate in the system. However, this usually presents analytical difficulties (digestion or liquefaction of solids, and/or very sophisticated analytical methods may be required). When the sorption equilibrium is established, the sorbate immobilized in the solid sorbent will be in equilibrium with the residual concentration of sorbate remaining in the liquid phase. Thus, it can easily be seen that the initial concentration of the sorbate (C_0) is of little relevance to the batch sorption-equilibrium tests. It can, perhaps, assist in identifying the final concentration range, which, of course, also depends on the amount of sorbent solids (X) in the system. It should also be noted that there is very little control over the value of C_{eq}. This value is subsequently used for calculating the uptake q from the mass balance on the sorbate in the system with solution volume V:

$$q = \frac{\left(C_0 - C_{eq}\right)V}{X} \tag{6.38}$$

The comparison of sorbent performance based on '% removal' is often encountered in the literature.

Models play a key role in transferring technologies from the laboratory to a full-scale application. Good models not only help in analyzing and interpreting experimental data but also in predicting the response of the systems to changing conditions. To describe the column breakthrough curves obtained at different bed heights, flow rates and inlet metal concentrations, different models are used. These include (Vijayaraghavan and Prabu, 2006),

- **Thomas model**

$$\frac{C}{C_0} = 1 + \exp\left(\frac{k_{TH}}{F}(Q_0 M - C_0 V_{eff})\right) \tag{6.39}$$

- **Yoon-Nelson model**

$$\frac{C}{C_0} = \frac{(\exp k_{YN} t - \tau k_{YN})}{1 + (\exp k_{YN} t - \tau k_{YN})} \tag{6.40}$$

- **Modified dose-response model**

$$\frac{C}{C_0} = 1 - \frac{1}{1 + \left(\frac{v_{eff}}{b_{mdr}}\right)^{a_{mdr}}} \tag{6.41}$$

where k_{TH} is the Thomas model rate constant ($mg^{-1}\ h^{-1}$), Q_0 the maximum solid-phase concentration of the solute ($mg\ g^{-1}$), V_{eff} the volume of metal solution passed into the column (l), k_{YN} the Yoon-Nelson model rate constant (min^{-1}), ô the time required for 50% adsorbate breakthrough (min) and a_{mdr} and b_{mdr} are the modified dose–response model constants.

- **Bed depth service time (BDST) model**

The performance of pilot-scale columns is analysed by plotting service time versus bed depth for several different flow ates and column lengths. Although the model provides a simple and comprehensive approach to running and evaluating pilot-scale tests, its validity is limited to the range of conditions used during the pilot tests. The bed depth service time (BDST) model, proposed by Bohart and Adams (1920) and subsequently modified by Hutchins (1973), is based on physically measuring the capacity of the bed at different breakthrough values. The simplified design model ignores the intraparticle mass transfer resistance such that the sorbate is sorbed onto the sorbent surface directly (Ko et al., 2000). The BDST model states that bed height (*Z*) and service time (*t*) of a column bears a linear relationship. The equation can be expressed as,

$$t = \frac{N_{0Z}}{C_0 v} + \frac{1}{K_a C_{.0}} \ln\left(\frac{C_0}{C_b} - 1\right) \tag{6.42}$$

where C_b is the breakthrough metal ion concentration (mg l^{-1}), N_0 is the sorption capacity of bed (mg l^{-1}), v is the linear velocity (cm h^{-1}) and K_a is the rate constant (l mg^{-1} h^{-1}).

6.4 RECENT DEVELOPMENTS

There is a need for fundamentally based modeling approach that integrates the principles of hydrodynamics, the irradiance field and photosynthesis to enhance biomass productivity by maximizing growth rate and efficiency in light use. There are many advances in this area as listed below. Modeling of kinetics in standard photobioreactors is difficult because of strong light gradients and lack of control of light energy. For the study of the microalgae kinetics and for modeling purposes a new kind of photobioreactor has been designed. (Csogor et al., 1999). Model based approaches are used to optimize illumination and to achieve higher light intensities and more uniform illumination. A novel design of gas induced mechanically agitated photobioreactors (GIMAP), using the principle of hallow shaft air, has been tested with *Spirulina* as a model (Kshirsagar et al., 2002). This is a special reactor having potential advantages like no separate power requirement for sparging. Moderate values of mass transfer coefficients and relatively low values of shear stress. Light availability inside the reactor is often the bottle neck in microalgal cultivation and for this reason much attention is being given to light limited growth kinetics of microalgae. The continuous culture studies are usually achieved using chemostat cultivations. Some researchers (Barbosa et al., 2003), have investigated the applicability of the new and dynamic cultivation method called acceleration stat (A-stat) to microalgae cultivation where light is the limiting substance. A-stat can be used as a fast and accurate tool to determine kinetic parameters and to optimize any specific type of photobioreactor. Photobioreactor sensing presents unique challenges, which are not met in conventional fermentors. First and foremost, photobioreactor processes are governed by photosynthesis, and hence many parameters important to photobioreactors are not considered in conventional fermentors. Development of effective model based estimator that is capable of tracking key cultures in stirred tank packed bed reactors has been reported (Li et al., 2003). This model is based on the extended Kalman filter in conjugation with online dissolved O_2 measurement. The growth can be studied by measurement of dissolved oxygen. In PBRs, the availability and the intensity of the light, which are affected by the cell movement, are the major factors controlling the biomass productivity. Hydrodynamics, hence, play a significant role in the reactor performance. Computer Automated Radioactive Particle Tracking (CARPT) technique has been employed to evaluate its feasibility for characterizing PBRs (Luo et al., 2003). Liquid velocity profile, cell movement and the temporal irradiance patterns, obtained by coupling the cell trajectories and irradiance distribution models have been determined.

Chapter 7

NUMERICAL PROBLEMS

Bioprocess technology is an interdisciplinary science combining life science and engineering. Therefore, it is essential to include numerical problems, which will enable the reader (especially students) to get a better feel of the subject. This Chapter presents some problems on microalgal cultivation, downstream processing and biosorption. Solutions are given for few numerical problems. This is followed by some problems suggested for self-study.

SOLVED PROBLEMS

Problem 1: Cultivation studies of *S. platensis* were carried out in a shake flask with rotation speed of 180 rpm at a light intensity of 250 Lux. The rate in change of biomass (r_x) was found out at different growth stage of the algae. Following results were obtained. (X: biomass concentration, g l^{-1})

X (g l^{-1})	0.06	0.13	0.19	0.20	0.30	0.38	0.39	0.72	0.56
r_x (g l^{-1} h^{-1})	0.0004	0.0008	0.0010	0.0012	0.0014	0.0016	0.0017	0.0019	0.0019

Calculate K_m and $r_{x,\ max}$

Solution:

Step 1

The growth rate of algae obeys Monod type law, which is given by

$$r_x = \frac{r_{x\ max}.X}{K_M + X}$$

Linearizing the equation,

$$\frac{1}{r_x} = \frac{K_m}{r_{x\ max}.X} + \frac{1}{r_{x\ max}}$$

This means that graph of $\frac{1}{r_x}$ vs. $\frac{1}{X}$ should give a straight line.

Step 2

Columns 3 and 4 are calculated and Fig. 7.1 is plotted.

X (g l^{-1})	r_x (g l^{-1} h^{-1})	1/X	1/ r_x
0.06	0.0004	16.82632	2397.938
0.13	0.0008	7.601944	1260.396
0.19	0.0010	5.397255	980.8266
0.20	0.0012	4.994656	829.1042
0.30	0.0014	3.308997	698.0754
0.38	0.0016	2.598109	629.9914
0.39	0.0017	2.536012	572.9255
0.72	0.0019	1.396462	537.728
0.56	0.0019	1.798691	20.2894

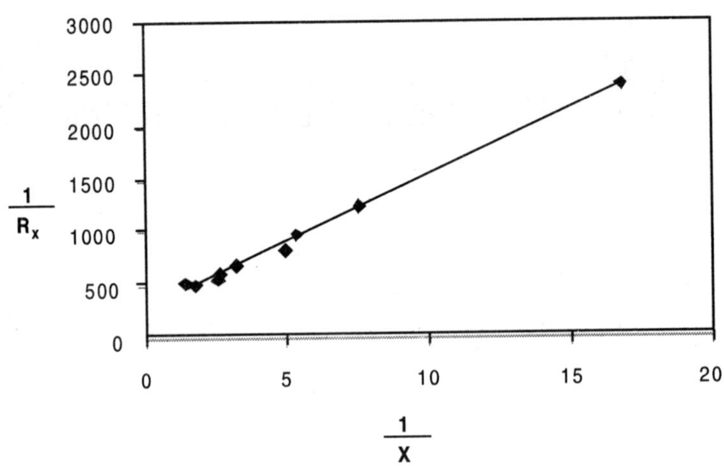

Fig. 7.1. Plot of $\frac{1}{r_x}$ vs. $\frac{1}{X}$

Step 3

Plot of $\frac{1}{r_x}$ vs. $\frac{1}{X}$ is a straight line. The y-intercept of the line is and slope of the line is

$$\text{y-intercept} = \frac{1}{r_{x\max}} = 294.23$$

$$\therefore \quad r_{x\max} = 0.00803 \text{ g l}^{-1}\text{ h}^{-1} \quad \textbf{Ans.}$$

$$\text{Slope} = \frac{K_m}{R_{x\max}} = 124.39$$

$$K_m = 2.36 \text{ g l}^{-1} \quad \textbf{Ans.}$$

Result : $K_m = 2.36$ g l^{-1} and $r_{x\max} = 0.00803$ g l^{-1} h^{-1}

C_0 (mg l^{-1})	Biomass loading (g/l)	C_{eq} (mg l^{-1})
25	0.01	11.3636
50	0.01	22.2727
75	0.01	49.5455
100	0.01	60.9091
125	0.01	73.1818
150	0.01	102.2727
175	0.01	119.5455
200	0.01	143.1818
225	0.01	160.9091

Problem 2. Biosorption experiments of the heavy metal, Cr (VI) by an alga *C. vulgaris* were performed in shake flasks. Shake flasks were rotated at a speed of 180 rpm at pH 1.5 and temperature 25°C. The data obtained are as follows:

State whether the data fits into Freundlich or Langmuir adsorption isotherm. Give the isotherm constants.

Solution:

Step 1

Langmuir adsorption isotherm should be assumed.

It is given by the equation,

$$q_{eq} = q_{\max} \frac{K_L.C_{eq}}{1+K_L C_{eq}}$$

Where q_{eq} is adsorbate loading at equilibrium (mg g^{-1}), C_{eq-} is the concentration of adsorbate in the fluid at equilibrium (mg l^{-1}), K_L is the Langmuir adsorption constant, and q_{max} is the maximum adsorbate loading to form a complete monolayer on the surface bound at high C_{eq}.

In the linearized form,

$$\frac{1}{q_{eq}} = \frac{1}{q_{eq}.K_L.C_{eq}} + \frac{1}{q_{max}}$$

This means that graph of $\frac{C_{eq}}{q_{eq}}$ vs. C_{eq} should give straight line.

Step 2

Columns 4 and 5 are calculated and Fig. 7.2 is plotted.

C_0 (mg l^{-1})	Biomass loading (g/l)	C_{eq} (mg l^{-1})	q_{eq} (mg g^{-1})	q_{eq}/C_{eq} (l g^{-1})
25	1	7.27	17.73	2.44
50	1	9.09	40.91	4.50
75	1	15.00	60.00	4.00
100	1	21.82	78.18	3.58
125	1	33.64	91.36	2.72
150	1	50.91	99.09	1.95
175	1	64.09	110.91	1.73
200	1	84.09	115.91	1.38
225	1	94.54	130.45	1.38
250	1	110.00	140.00	1.27

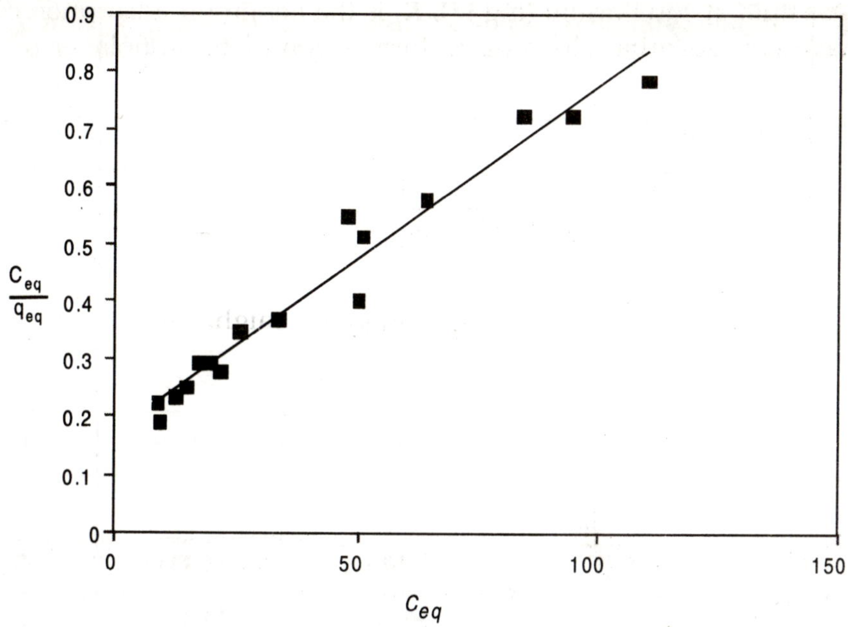

Fig. 7.2. $\frac{C_{eq}}{q_{eq}}$ vs C_{eq}

From the graph it can be observed that the data is not fitting well to a straight line so we must assume another adsorption isotherm.

Step 3

Freundlich adsorption isotherm is assumed

It is given by the relation,

$$q_{eq} = K_F \, C_{eq}n$$

where q_{eq} is adsorbate loading at equilibrium (mg g^{-1}), C_{eq} is the concentration of adsorbate in the fluid at equilibrium (mg l^{-1}), and K_F is the Freundlich adsorption constant.

In linearized form,

$$\ln \; q_{eq} = K_F + n \times \ln C_{eq}$$

This means that graph of ln q_{eq} vs. ln C_{eq} should give straight line.

C_0 (mg l^{-1})	Biomass loading (g/l)	C_{eq} (mg l^{-1})	q_{eq} (mg g^{-1})	ln q_{eq}	ln
25	1	7.27	17.73	1.98	2.88
50	1	9.09	40.91	2.21	3.71
75	1	15.00	60.00	2.71	4.09
100	1	21.82	78.18	3.08	4.36
125	1	33.64	91.36	3.52	4.51
150	1	50.91	99.09	3.93	4.60
175	1	64.09	110.91	4.16	4.71
200	1	84.09	115.91	4.43	4.75
225	1	94.54	130.45	4.55	4.87
250	1	110.00	140.00	4.70	4.94

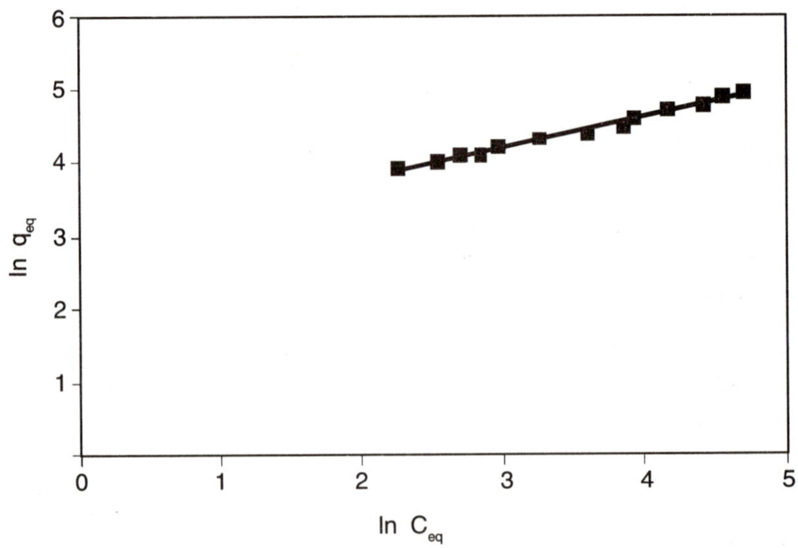

Fig. 7.3. *Plot of ln q_{eq} vs. ln C_{eq}*

Step 4

Columns 5 and 6 are calculated and Fig. 7.3 is plotted.

Plot of ln q_{eq} vs. ln C_{eq} is a straight line, which shows that the data fits the Freundlich adsorption isotherm. The *y*-intercept of the line is ln K_F and slope of the line is n.

$$y\text{-intercept} = \ln K_F = 2.93$$

$$\therefore \quad K_F = 18.90 \qquad \textbf{Ans.}$$

$$\text{Slope} = n = 0.42 \qquad \textbf{Ans.}$$

Result: The biosorption data fits the Freundlich adsorption isotherm. The Freundlich isotherm constants, K_F = 18.90 and n = 0.42.

Problem 3. The alga *S. platensis* was immobilized in calcium alginate gel beads of size 2 mm. These gel beads were packed in a column to adsorb Cr(VI) metal ions. Column parameters are as follows:

Height of column: 15 cm; Flow rate: 3.5 ml/min; C_0 = 100 mg/l

t (min)	C (mg/l)
0	0.75
3	1.50
6	2.24
9	3.15
12	3.45
15	4.65
18	5.41
21	6.33
24	7.70
27	10.30
30	20.47
33	30.82
36	39.50
39	46.16
42	55.00
45	62.00
48	68.00
51	75.00

Contd...

54	80.00
57	85.00
57	85.00
60	89.00
65	93.00
70	97.00
75	100.00
80	100.00
85	100.00
90	100.00

For immobilized alga column, the concentration of Cr(VI) at different time interval is given below:

Calculate the Total metal removal (%).

Solution:

The total quantity of metal ions adsorbed in the column (W_{ad}) is calculated from the area over the breakthrough curve (outlet metal concentration vs. time) multiplied by flow rate. Dividing W_{ad} by biosorbent mass (*M*) leads to the uptake capacity (*q*) of the biomass. The total amount of the metal sent to the column can be calculated from the equation:

$$W = \frac{C_0 . F . t_e}{1000}$$

where C_0 is the inlet metal ion concentration (mg l^{-1}), F the volumetric flow rate (ml h^{-1}) and t_e is the exhaustion time (h).

Total metal removal (%) can be calculated from equation,

$$\text{Total metal removal (\%)} = \frac{W_{ad}}{W} \times 100$$

Step 1

Breakthrough curve i.e. C/C_0 vs. t is plotted.

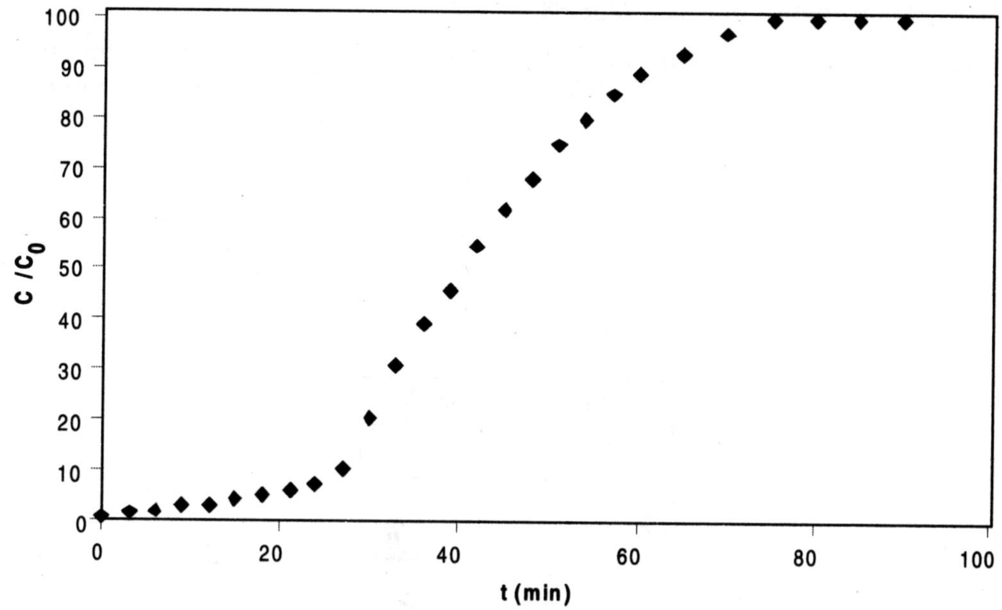

Area above the curve is 4251.01 units.

Step 2

The total amount of the metal sent to the column can be calculated from the equation:

$$W = \frac{C_0.F.t_e}{1000}$$

$$= \frac{100 \times 3.5 \times 75}{1000} W = \frac{C_0.F.t_e}{1000}$$

$$= 26.25$$

$$W_{ad} = \frac{\text{area above curve} \times \text{flowrate}}{1000}$$

$$= \frac{4251.01 \times 3.5}{1000}$$

$$= 14.87 \text{ mg}$$

Step 3

Total metal removal (%) can be calculated from equation,

$$\text{Total metal removal (\%)} = \frac{W_{ad}}{W} \times 100$$

$$= \frac{26.25}{14.87} \times 100$$

$$= 56.58\% \quad \textbf{Ans.}$$

Result: Total metal removal is 56.68%.

Problem 4. Column studies were carried on the adsorption of Cr(VI) ions on immobilized *S. platensis.* The alga was immobilized in Ca-alginate gel beads. The column parameters are as follows:

Bed height: 28 cm, flow rate: 3.5 ml min^{-1} and initial metal ion concentration: 100 mg l^{-1}.

Calculate the Number of transfer units (NTU) and Height equivalent to theoretical plate (HETP) for the column

t (min)	C (mg/l)
0	0.15
3	2.40
6	3.15
9	3.90
12	4.50
15	4.81
18	5.87
21	6.63
24	6.93
27	7.24
30	7.85
33	8.15
36	9.07
39	10.61
42	16.84
45	25.00
48	33.00
51	40.00
54	46.00
57	51.00
60	55.90
63	60.80
66	65.50

Contd...

69	70.20
72	75.00
75	79.00
78	82.00
81	85.00
84	88.00
87	90.00
90	91.50
93	93.00
96	94.50
99	95.80
102	96.50
105	97.30
108	98.00
111	99.00
114	100.00
117	100.00
120	100.00
125	100.00

Solution:

Step 1

Breakthrough curve, C/C_0 vs. t is plotted:

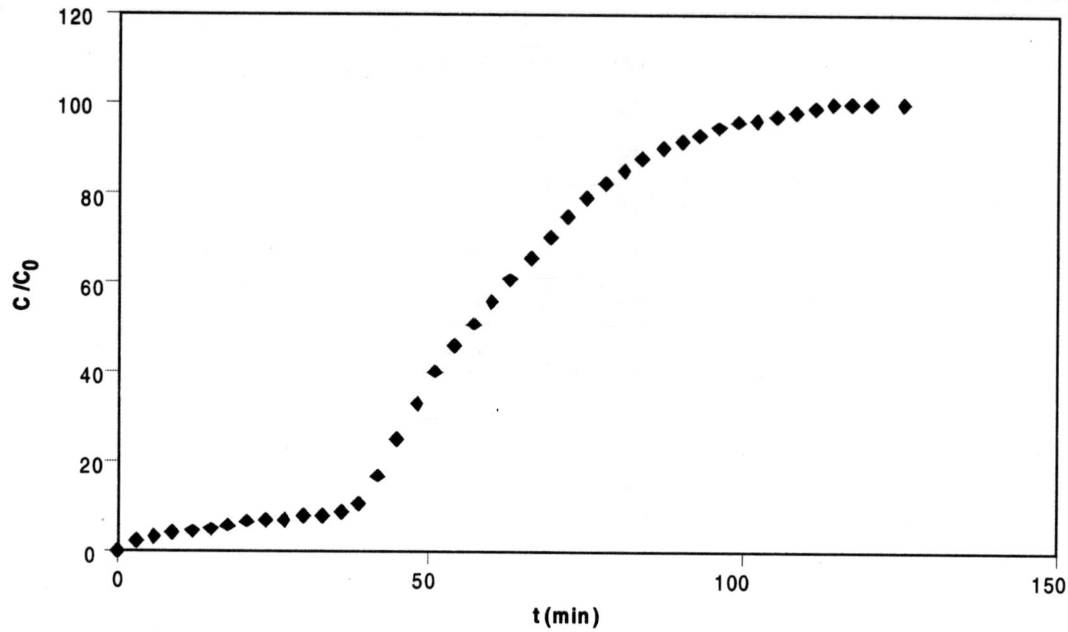

Area under the curve of breakthrough curve (AUC) = 5746 mg min l^{-1}.

Step 2

Total metal ions loaded during breakthrough (W)

$W = C_0\,(t_e - t_b) = 100\,(114 - 39) = 7500$ mg min l^{-1}

Metal ions adsorbed during breakthrough (W_a)

$$W_a = (W - \text{AUC})$$
$$= (7500 - 5746)$$
$$= 1754 \text{ mg min } l^{-1}$$

Fraction of metal ions adsorbed

$$f = \frac{W_a}{W}$$
$$= \frac{1754}{7500}$$
$$= 0.234$$

Step 3

Length of adsorption zone (Z_a)

$$Z_a = \frac{Z(V_e - V_b)}{V_e - (1 - f)(V_e V_b)}$$

$$= \frac{28.(0.399 \times 0.1365)}{0.399 - (1 - 0.234)(0.399 \times 0.1356)}$$

$$= 37.13 \text{ cm}$$

Step 4

Number of transfer unit (*N*)

$$N = 16 \left(\frac{V_R}{V_E - V_B} \right)^2$$

$$= 16 \left(\frac{0.267}{0.399 - 0.1365} \right)^2$$

$$= 16.64\ H''\ 17 \quad \textbf{Ans.}$$

Height equivalent to theoretical plate (HETP)

$$\text{HETP} = \frac{Z_a}{N}$$

$$= \frac{37.13}{17}$$

$$= 2.18 \text{ cm} \quad \textbf{Ans.}$$

Result: for the given column, NTU = 17 and HETP = 2.18 cm.

PROBLEMS FOR SELF STUDY

1. *C. vulgaris* was cultivated in a bench scale fermentor with an agitation speed of 150 rpm and at a light intensity of 1400 Lux. The rate in change of biomass (r_x) of biomass concentration (X, g l^{-1} is given in the following table:

Calculate K_m and $r_{x,\ max}$ (**Hint:** See solved problem 1)

X (g l^{-1})	0.09	0.18	0.27	0.36	0.41	0.49	0.58	0.57	0.70
r_x(g l^{-1} h^{-1})	0.0011	0.0017	0.0027	0.0022	0.0052	0.0039	0.0049	0.0097	0.0056

(**Ans.** K_m = 1.02 g l^{-1}; $r_{x\ max}$= 0.013 g l^{-1} h^{-1})

2. Cultivation studies of the alga *C. sorokiniana* was carried out in a shake flask at 180 rpm and light intensity of 3200 Lux. The biomass of the alga (X, g l^{-1}) was found out at different time intervals (t, h). The following results were obtained:

Calculate K_m and $r_{x,\ max}$. (**Hint:** See solved problem 1)

t (h)	24	48	72	96	120	144	168	192	216
X (g l^{-1})	0.08	0.15	0.22	0.29	0.37	0.44	0.52	0.58	0.65

(**Ans.** K_m = 1.24 g l^{-1}; $r_{x\ max}$= 0.0303 g l^{-1} h^{-1})

3. The cultivation studies of algae, *S. platensis,* in a shake flask at different light intensities were carried out in order to find out relation between growth rate and light intensity at a constant temperature of 25°C. The growth rate (r_x) at different light intensities (I, Lux) and biomass concentration (X, g l^{-1}) is given in the following table:

Growth rate (g l^{-1} h^{-1}) at biomass concentration (X, g l^{-1})

Light intensity (Lux)	X= 0.05	0.1	0.2	0.3	0.4	0.5	0.6	0.7	0.8
250	0.00014	0.00028	0.00056	0.00084	0.00112	0.0014	0.00168	0.00196	0.00224
800	0.0003	0.0007	0.0014	0.0020	0.0027	0.0034	0.0041	0.0048	0.0055
1400	0.0005	0.0009	0.0019	0.0028	0.0038	0.0047	0.0057	0.0066	0.0076
2000	0.0006	0.0011	0.0022	0.0034	0.0045	0.0056	0.0067	0.0079	0.0090
2600	0.0006	0.0012	0.0025	0.0037	0.0050	0.0062	0.0075	0.0087	0.0100
3200	0.0007	0.0013	0.0027	0.0040	0.0054	0.0067	0.0080	0.0094	0.0107

Using the growth model,

$$r_x = \mu.X$$

$$X\mu = \frac{\mu_{max}.1}{K_I + I}$$

Find out μ_{max} and K_I. (**Hint:** See solved problem 1 and section 3.2)

(**Ans.** μ_{max} = 0.0197 h^{-1}; K_I = 1508.52 Lux)

4. An aliquot of *S. platensis* cells were suspended in acetone and vortex mixed until a white precipitate appeared and b-carotene along with chlorophyll is extracted. Explain the mass transfer resistances in the process.

5. In the biosorption of the metal ions by algae, following data was obtained.

C_0 (mg l^{-1})	Biomass loading (g/l)	C_{eq} (mg l^{-1})
100	2	45.82
100	4	20.16
100	6	8.31
150	2	123.99
150	4	119.53
150	6	96.20
200	2	196.70
200	4	83.95
200	6	63.11

Which adsorption isotherm does the data fit into? Give the isotherm constants.

(**Hint:** See solved problem 2)

(**Ans.** Freundlich adsorption isotherm; $K_F = 9.14$; $n = 0.25$)

6. The alga *C. vulgaris* was immobilized in calcium alginate gel beads of size 3 mm. These gel beads are then packed in a column to adsorb copper ions. Column parameters are as follows:

Height of column: 28 cm; Flow rate: 3.5 ml/min; $C_0^- = 200$ mg/l

For the immobilized alga column, concentration of Cu (V) at different time interval were as follows:

t (min)	C (mg/l)
0	5.56
3	6.32
6	7.54
9	9.53
12	9.69
15	9.84
18	10.77
21	12.31
24	18.73
27	29.01
30	43.58
33	52.00
36	63.00
39	70.00
42	76.00
45	83.00
48	88.00
51	91.00
54	94.00
57	96.00
60	97.00
63	98.00
66	99.00
69	100.00
72	100.00

Calculate Total (%) metal removal. (**Hint:** See solved problem 3)

(**Ans.** 49.59%)

7. During adsorption experiments of Cr (VI) ions on immobilized *S. platensis* in a packed column, the following data was obtained.

Column parameters were

Bed height: 28 cm, flow rate: 3.5 ml min^{-1} and initial metal ion concentration: 150 mg l^{-1}.

t (min)	C (mg/l)
0	2.30
3	3.75
6	4.65
9	5.56
12	6.32
15	7.54
18	9.53
21	9.69
24	9.84
24	9.84
27	10.77
30	12.31
33	18.73
36	29.01
39	43.58
42	52.00
45	58.00
48	65.00
51	72.00
54	78.00
57	84.00
60	88.00
63	91.00
66	93.00

Contd...

69	95.00
72	96.00
75	97.00
78	98.00
81	99.00
84	100.00
87	100.00
90	100.00

Calculate NTU and HETP for the column. (**Hint:** See solved problem 4)

(**Ans.** NTU = 18; HETP = 1.93 cm)

8. For the following data, obtained in the biosorption experiments in packed bed column, find out the BDST model parameters and comment on them.

Bead height (cm)	Service time (exhaust time)
15	75
22	100
28	114

Flow rate: 3.5 ml/min; column diameter: 2.5 cm; C_0: 100 mg/l; C_b: 8 mg/l

(**Hint:** See introduction BDST model)

(**Ans.** *NO*= 128.98 mg/*L*)

NOMENCLATURE

A	local absorption rate ($W.m^{-3}$)
A_λ	local absorption rate for radiation of wavelength λ ($W.m^{-4}$)
a, b_{mdr}	modified dose-response model constants
C	concentration of Cr(VI) ion in the fluid at any time t (mg l^{-1})
C_0	initial/inlet metal ion concentration (mg l^{-1})
C_b	breakthrough metal ion concentration (mg l^{-1})
C_x	biomass concentration ($kg.m^{-3}$)
Di	diffusion coefficient ($m^2.s^{-1}$)
E_a	absorption mass coefficient ($m^2.kg^{-1}$)
F	hyperbolically transformed nutrient concentration = $x/(K+x)$
F	radiant energy flux ($W.m^{-2}$)
F	volumetric flow rate (ml h^{-1})
F_λ	radiant energy flux for the radiation of wavelength λ ($W.m^{-3}$)
I_λ	specific intensity for the radiation of wavelength λ ($W.m^{-3}$)
J	mean intensity ($W.m^{-2}$)
J_λ	mean intensity for the radiation of wavelength λ ($W.m^{-3}$)
J_c	mean diffusion rate ($mol.m^2.s^{-1}$)
K_a	apparent Michaelis constant for bicarbonate transfer
K_a	rate constant for adsorption of metal ions in the column (l mg^{-1} h^{-1})
K_F	Freundlich adsorption constant
K_j	half saturation constant for total energy available ($W.m^{-2}$)
K_L	Langmuir adsorption constant (mg l^{-1})
k_{TH}	Thomas model rate constant (mg/h)
k_{YN}	Yoon–Nelson model rate constant (min^{-1})
L	length of reactor (m)

I	light energy ($W.m^{-2}$)
I_{opt}	light energy ($W.m^{-2}$)
M	biosorbent mass (g)
N_0	sorption capacity of bed ($mg\ l^{-1}$)
P_i	permeability constant
q	mass of metal ion adsorbed per unit mass of original adsorbent at any time ($mg\ g^{-1}$)
q_{eq}	mass of metal ion adsorbed per unit mass of original adsorbent at equilibrium ($mg\ g^{-1}$)
q_{max}	maximum adsorbate loading ($mg\ g^{-1}$)
Q_o	maximum solid phase concentration of solute ($mg\ g^{-1}$)
R	respiration rate ($mol.s^{-1}$)
R_i	inner radius (cm)
r_r	growth rate of microalgae ($g\ l^{-1}\ h^{-1}$)
T	temperature (℃)
t	time (s)
v	linear velocity of solution in the column ($cm\ h^{-1}$)
V	volume (m^3)
V_a	active transport rate ($mol.s^{-1}$)
V_c	CO_2 fixation rate ($mol.s^{-1}$)
V_{Cn}	inorganic carbon uptake rate in presence of nitrate ($mol.s^{-1}$)
V_{Co}	inorganic carbon uptake rate in absence of nitrate ($mol.s^{-1}$)
V_{eff}	volume of metal solution passed into the column (l)
V_p	transformation rate ($mol.s^{-1}$)
V_s	CO_2 assimilation rate ($mol.s^{-1}$)
x	constant = I/I_{opt}, nutrient concentration in environment (mol)
Z	column height (cm)
Z	dimensionless length = z/L.
z	length (m)
Z_a	length of adsorption zone (cm)

Greek letter

α	$[E_a/(E_a+E_s]^{1/2}$, dimensionless
β	sharpness coefficient, constant
δ	(E_a+E_s). α. $C_x.L$, dimensionless

ϕ,θ angle (rad)

θ_0 temperature term (C)

θ_{Opt} optimum temperature

¼ specific growth rate (h^{-1})

$¼_{max}$ maximum specific growth rate (h^{-1})

Ä time required for 50% adsorbate breakthrough (min)

REFERENCES

Abel W.O., Knebel W., Koop H.U., (1989). A cytokinin-sensitive mutant of the moss *Phycomitrella patens,* defective in the chloroplast division. *Protoplasma* 152: 1–3.

Aiba, S. (1982). "Growth kinetics of photosynthetic microorganisms". *Advances in Biochemical Engineering,* 23: 85–156.

Akhtar, N., Iqbal, J., Iqbal, M. (2004). "Removal and recovery of nickel (II) from aqueous solution by loofa sponge-immobilized biomass of *Chlorella sorokiniana*: characterization studies". *Journal of Hazardous Materials,* B108: 85–94.

Aksu Z., Açkel Ü., Kutsal T. (1997). Application of multicomponent adsorption isotherms to simultaneous biosorption of iron(III) and chromium(VI) on C. vulgaris. J Chem Technol Biotechnol 70: 368–78.

Aksu, Z. (2002). "Determination of the equilibrium, kinetic and thermodynamic parameters of the batch biosorption of nickel (II) ions onto *Chlorella vulgaris" Process Biochemistry,* 38: 89–99.

Al-Qunaibit, M., Khalil, M., Al-Wassil, A. (2005). "The effect of solvents on metal ion adsorption by the alga Chlorella vulgaris". *Chemosphere,* 60: 412–418.

Anatoly, G., Hu, Q., Richmond, A., (1996). Photic Volume in Photobioreactors Supporting Ultra high Population Densities of the Photoautotroph *Spirulina platensis.* App. Environ Microbial. pp. 1570–1573.

Apt K.E., Behrens P.W. (1999). Commercial Developments in Microalgal Biotechnology, 35, 215–226.

Apt K.F., Kroth-Pancic P., Grossman A.K. (1996). Stable nuclear transformation of the diatom *Phoeodactylum tricornutum. Mol Gen Genet* 252: 572–579.

Augugliaro, V., Loddo, V., Marci, G., Palmisano, L., López-Muñoz, M.J. (1997). "Photocatalytic oxidation of cyanides in aqueous titanium dioxide suspension". *Journal of Catalysis,* 166: 272–283.

Babcock Jr, R.W., Malda, J., Radway, J.C., 2002. Hydrodynamics and mass transfer in a tubular airlift photobioreactor. J. Appl Phycol. 14, pp. 169–184.

Barbosa, M.J., Albrecht, M., Wijffels, R. (2003b). Hydrodynamic stress and lethal events in sparged microalgae cultures. Biotechnol Bioeng. 83, pp. 112–120.

Barbosa, M.J., Janssen, M., Nienke, H., Tramper, J., Rene, H., Wijffels, R.H., (2003). Microalgae Cultivation in Air-Lift Reactors: Modeling Biomass Yield and Growth Rate as a Function of Mixing Frequency. Biotechnol. and Bioeng. 82, (2), pp. 170-179.

Barclay, W., Zeller, S. (1996). Nutritional enhancement of n-6 fatty acids in rotifers and *Artemia nauplii* by feeding spray dried *Schizpchytrium* sp. Journal of World Aquaculture Society, 27, 314–322.

Behrens, P.W., Kyle, D.J. (1996). Microalgae as a source of fatty acids. Journal of Food Lipids, 3, 259–272.

Belay, A., (1997). Mass culture of *Spirulina* outdoors—The Earthrise Farms experience. In: Vonshak, A. (Ed.), *Spirulina platensis* (*Arthrospira*): Physiology, cell-biology and biotechnology. Taylor and Francis, London, pp. 131–158.

Belter, P.A., Cussler, E.L., Hu, W.S. (1988). Bioseparations: Downstream Processing for Biotechnology. New York: Wiley.

Ben-Amotz, A., (1995). New mode of *Dunaliella* biotechnology: two-phase growth for â-carotene production. J Appl Phycol. 7, pp. 65–68.

Ben-Amotz, A., Avron, M. (1987). The biotechnology of mass culturing of Dunaliella for products of commercial interest. In: CresswellRC, Rees TAV, Shah, N., editors. Algal and cyanobacterial technology. London: Longman, pp. 90–114.

Benemann, J.R. (1979). "Production of nitrogen fertilizer with nitrogen-fixing blue - green algae". *Enzyme and Microbial Technology,* 1 (2): 83–90.

Bermejo, R.R., Talavera, E.M., Alvarez-Pez, J.M. (2001). Chromatographic purification and characterization of b-phycoerythrin from *Porphyridium cruentum.* Semi preparative HPLC separation and characterization of its subunits. J. Chromatography. 917, pp. 135–45.

Blanchemain, A., Grizeau, D. (1999). Increased production of eicosapentaenoic acid by *Skeletonema costatum* cells after decantation at low temperature. Biotechnol Tech. 13, pp. 497–501.

Bohart, G. and Adams, E.Q. (1920). "Some aspects of the behavior of charcoal with respect to chlorine". *Journal of Americal Chemical Society,* 42: 523–544.

Bold, H.C., Wynne, M.J. (1978). Introduction to the algae. Structure and reproduction: Englewood Cliffs. New Jersey, Prentice-Hall, Inc., pp. 706.

Borowitzka MA (1992). Algal biotechnology prodcts an processes-matching science and economics Journal of Applied Phycology, 4, 267–279.

Boyd MR, Gustafson KR, McMahon JB, Shoemaker RH, O Keefe BR, Mori T, Gulakowski RJ, L Wu, Rivera MI, Laurencot CM, Currens MJ, Cardellina JH, Buckheit Jr, RW, Nara PL, Pannell LK, Sowder II RC, Henderson LE (1997). Discovery of Cyanovirin N, a novel immunoidefeciency virus inactivating protein that binds viral surface envelope glycoprotein gp 120: potential applications to microbicide development. Antimicrobial agents and Chemotherapy, 41, 1521–1530.

Brierley, J.A., Brierley, C.L., Goyak, G.M. (1986). In *Fundamental and Applied Biohydrometallurgy,* Lawrence, R.W., Branion, R. M.R., Ebner, H.G., (Eds), (pp. 291–304), Elsevier. cf. Kratochvil and Volesky (1998).

Brower V (1998) Nutraceuticals: poised for a healthy slice of the health care market? Nature Biotechnology, 16, 728–731.

Brown L.M. and Zeiler, K.G. (1993). "Aquatic biomass and carbon dioxide trapping". *Energy conservation and management,* 34(9-11): 1–15.

Buwa, V.V., Ranade, V.V. (2002). Dynamics of gas-liquid flow in a rectangular bubble column: experiments and single/multi-group CFD simulations. Chem Eng Sci. 27, pp 4715–4736.

Cannell, R.J.P., (1990). Algal biotechnology. Appl Biochem Biotechnol. 26, pp. 85–105.

Carlozzi, P; Torzillo, G., (1996). Productivity of Spirulina in a strongly curved outdoor tubular photobioreactor. Appl Microbiol Biotechnol. 45, pp. 18-23

Cartens, M., Molina, E., Robles, A., Gimenez, A., Ibanez, M.J. (1996). Eicosapentaenoic acid (20:4n-3) from the marine microalga *Phaeodactylum tricornutum.* J Am Oil Chem Soc. 73, pp. 1025–31.

Cerutti H, Johnson AM, Gillham NW et al. (1997). Epigenetic silencing of a foreign gene in nuclear transformants of Chlamydomonas. *Plant Cell* 9: 925–945.

Chapman, V.J. (1970). Seaweed and Their Uses. Second ed. Methuen & Co. Lt., London.

Chaumont, D. (1993). Biotechnology of algal biomass production: A review of systems for outdoor mass culture. J. Appl. Phycol. 5, pp. 593–604.

Cheney PD, Metz B, Stiller J. *(2001). Agrobacterium*-mediated genetic transformation in the macroscopic marine alga *Porphyra yezoensis. J Phycol* 37: (suppl.) p. 11.

Chisti, Y., (1989). Airlift bioreactors. New York: Elsevier, p. 355.

Chisti, Y., (1998). Pneumatically Agitated Bioreactors in Industrial and Environmental Bioprocessing:Hydrodynamics, Hydraulics and Transport Phenomena. Appl.Mech.Rev. 51, p. 33.

Chisti, Y., (1999a)."Mass Transfer," Encycleopedia of Bioprocess Technology. Fermentation, Biocatalysis and bioseperation.Vol. 3, Flickinger, M.C., Drew, S.W., (Eds.), Wiley, New York, p. 1607.

Chisti, Y. (1999b). "Shear Sensitivity,"Encylopedia of Bioprocess Technology. Fermentation, Biocatalysis and Bioseperation.Vol. 5, Flickinger, M.C., Drew, S.W., (Eds.), Wiley, New York, p. 2379.

Chisti, Y., Moo-Young, M. (1986). Disruption of microbial cells for intracellular products. Enzyme Microb Technol. 8, pp. 194–204.

Chisti, Y., Moo-Young, M., (1987). Airlift reactors: Characteristics, Applications and Design Considerations. Chem. Eng. Commun. 60, p. 195.

Chisti, Y., Moo-Young, M., (1989). On the Calculation of Shear Rate and Apparent Viscosity in Airlift and Bubble Column Bioreactors. Biotechnol. Bioeng. 34, p. 1391.

Chojnacka, K., Chojnacki, A., Gorecka, H. (2005). "Biosorption of Cr^{3+}, Cd^{2+} and Cu^{2+} ions by blue–green algae Spirulina sp.: kinetics, equilibrium and the mechanism of the process". *Chemosphere,* 59: 75–84.

Chong, K.H. and Volesky, B. (1995). "Description of two-metal biosorption equilibria by Langmuir-type models". *Biotechnology and Bioengineering,* 47 (4): 451–460.

Chow KC, Tung WL. (1999). Electro transformation of Chlorella vulgaris. Plant Cell Reports 18: 778–780.

Ciferri, O., (1983). *Spirulina,* the Edible Microorganism, Microbiological Reviews. pp. 551-578.

Cohen Z, Norman HA, Heimer YM (1995). Microalgae as a source of omega-3-fatty acids. World Rev. Nutr Diet, 77, 1–31.

Collos, Y., (1997). "A physiological basis for estimating inorganic carbon release during photosynthesis by natural phytoplankton"., Ecological Modeling.*,* 96, 285–292, (1997).

Cornet, J.F., Dussap, C.G. and Dubertret, G., (1992). "A structural model for simulation of cultures of cyanobacteria *Spirulina platensis* in a photobioreactor: Coupling between light transfer and growth kinetics", Biotechnology and Bioengineering, 40, 817–825.

Crist, R.H., Oberholser, K., Shank, N. Nguyen, M. (1981). "Nature of bonding between metallic ions and algal cell walls". *Environmental Science and Technology,* 15: 1212–1217.

Csogor, Z., Harrenbaur, M., Perner, I. and Posten, C. (1999). "Design of photobioreactor for modeling purposes", Chem. Eng. and Processing, 38, 517–523.

Dakiky, M., Khamis, M., Manassra, A., Mer'eb, M. (2002). "Selective adsorption of Chromium (VI) in industrial wastewater using low cost abundantly available adsorbents". *Advances in Environmental Research,* 6: 533–540.

Danesi, E.D.G., Rangel-Yagui, C.deO., de Carvalho, J.C.M., Sato, S., 2002. An investigation of replacing nitrate by urea in the growth and production of chlorophyll by *Spirulina platensis.* Biomass and Bioenergy. 23, pp. 261–269.

Daniel de Alava, Correa de Mello, P., Wagener, K ., (1997). The Relevence of the CO_2 Partial Pressure of Sodium Bicarbonate Solutions for the Mass Cultivation of the Microalgae *Spirulina.* J. Braz. Chem. Soc. 8, (5), pp 447–450.

Darnall, D.W. (1991). *Innovative Haz. Waste Treat. Technol., Ser. Biol. Proc.* 3: 65–72. cf. Kratochvil and Volesky (1998).

Davis, E. A., Dedrick, J., French, C.S., Milner, H.W., Myers, J., Smith, J. H. C., Spoehr, H. A. (1953). Laboratory experiments on *Chlorella* culture at the Carnegie Institution of Washington, Department of Plant Biology, pp. 105–153. In: Burlew, J.S., (ed.), Algal culture from laboratory to pilot plant. Carnegie Institution of Washington Publication No. 600. Carnegie Institution, Washington, DC.

Day, J.G., Benson, E.E., Fleck, R.A. (1999). "In vitro culture and conservation of microalgae: applications for aquaculture, biotechnology and environmental research". *In Vitro Cell. Dev.–Plant,* 35: 127–136.

Deckwer, W.D,. (1992). Bubble Column Reactors. Wiley, New York.

Dellert SF, Nowicki MJ, Farrell MK, Delente J, Fuebi JE (1997). The ^{13}C-Xylose breathe test For the diagnosis of small bowel bacterial overgrowth in children. Journal of Pediatric Gastroenterology Nutrition, 25, 153–158.

Dodd, J.C., (1979). Algae production and harvesting from animal wastewaters. Agric Wastes. 1, pp. 23–37.

Dönmez, G.C., Aksu, Z., Öztürk, A., Kutsal, T. (1999). "A comparative study on heavy metal biosorption characteristics of some algae". *Process Biochemistry,* 34: 885–892.

Doucha, J., Livansky, K., 1995. Novel outdoor thin-layer high densitymicroalgal culture system: Productivity and operational parameters. Algol. Stud. (Trebon) 76, pp. 129–147.

Dufosse L, Galaup P, Yaron A, Arad Shoshana M, Blanc P, Murthy K, Ravishankar GA (2005). Microorganisms and microalgae as sources of pigments for food use: a scientific oddity or an industrial reality? Trends in Food Science and Technology 16, 389–406.

Dunahay TG, Jaevis EE, Dias SS et al. Manipulation of microalgal lipid production using genetic engineering. *Appl Biochem Biotechnol* 58: 223–231, 1996.

Dunahay TG, Jarvis EE, Roessler PG (1995). Genetic transformation of the diatoms *Cyclotella cryptica* and *Naviculla saprophila.* Journal of Phycology 31, 1004–1012.

Dunstan, G.A, Vokman, J.K., Jeffrey, S.W., Barret, S.M., (1992). Biochemical composition of microalgae from the green algal classes Chlorophyceae and Parasinophyceae: 2. Lipid classes and fatty acid. J.Exp Mar Biol Ecol. 161, pp. 115–34.

Eilers P. H.C and Peeters J. C. H. A model for the relationship between light intensity and the rate of photosynthesis in phytoplankton' Ecological Modelling, Volume 42, Issues 3-4, September 1988, pp. 199–215.

Eisenberg, D.M., Oswald,W.J., Benemann, J.R., Goebel, R.P., Tiburzi, T.T. (1981). "Methane fermentation of microalgae". *Anaerobic Dig. Proc. Int. Symp.* 1st, 99–111.

Elmaleh, S., Coma, J., Grasmick, A., Bourgade, L., (1991). Magnesium induced algal flocculation in a fluidized bed. Water Sci Technol. 23, pp. 1695–702.

Endo, H., Hosoya, H., Koibuchi, T., (1977). Growth yields of *Chorella regularis* in dark-heterotrophic continuous culture using acetate. J. Ferment and Technol.

Esposito A, Pagnanelli F, Veglio F (2002). pH-related equilibria models for biosorption in singlemetal systems. Chem Eng Sci. 57:307–13.

Ferraz AI, Tavares T, Teixeira JA. Cr(III) removal and recovery from Saccharomyces cerevisiae. Chem. Eng. J. 2004; 105: 11–20.

Figueira MM, Volesky B, Ciminelli VST, Roddick FA. Biosorption of metals in brown seaweed biomass. Water Res 2000; 34 (1): 196–204.

Fischer H, Robl I, Sumper M et al. (1999). Targeting and covalent modification of cell wall and membrane proteins heterologously expressed in the diatom *Cylindrotheca fusiformis* (Bacillariophyceae). *J Phycol* 35: 113–120.

Fourest, E. and Roux, J.C. (1992). *FEMS Microbiol. Rev.* 14: 325–332. Kratochvil and Volesky (1998).

Fourest, E. and Volesky, B. (1996). "Contribution of Sulfonate Groups and Alginate to Heavy Metal Biosorption by the Dry Biomass of *Sargassum fluitans*". *Environmental Science and Technology,* 30 (1): 277–282.

Freundlich, H. (1907) Z. Phys. Chem. 57, 385–470. cf. Kratochvil and Volesky (1998).

Fridlyand,L.E. (1997) "Models of CO_2 concentrating mechanism in microalgae taking into account cell and chloroplast structure", Biosystems, 44, 41-57.

Geil KM, Mandoli DF (1996). Development of mass transformation in *Actabularia acetabulum. J Phycol* 32: (suppl.) 17–18.

Giersch, C. (2003). "Stationary diffusion gradients associated with photosynthetic carbon flux- a study of compartmental versus diffusion reaction models", J. Theor. Biol., 224, 385-397.

Gill I, Valivety R (1997). Polyunsaturated fatty acids, Part I: occurrence, biological activities and applications. Trends in Biotechnology, 15, 401–409.

Gimenez Gimenez, A., Ibanez, M.J., Robles, A., Molina, E., Garcýa, S., Esteban, L. (1998). Downstream processing and purification of eicosapentaenoic (20:5 n-3) and arachidonic acids (20:4 n-6) from the microalga *Porphyridium cruentum.* Bioseparation. 7, pp. 89–99.

Glazer AN (1994). Phycobiliproteins— a family of valuable, widely used fluorophores. Journal of Applied Phycology, 6, 105–112.

Golueke, C.G., Oswald, W.J. (1965). Harvesting and processing sewage grown planktonic algae. J Water Pollut Control Fed.37, pp. 471–98.

Grobbelaar, J.U; 1991. The influence of light/dark cycles in mixed algal cultures on their productivity. Bioresource Technol. 38, pp. 189–194.

Grobbelaar, J.U., Soeder, C.J., Groeneweg, E.S., Hartig, P. (1988). "Rates of biogenic oxygen production in mass-cultures of microalgae, absorption of atmospheric oxygen and oxygen availability for waste-water treatment". *Water Research,* 22: 1459–1464.

Gudin, C., Chaumont, D., (1983). Solar biotechnology study and development of tubular solar receptors for controlled production of photosynthetic cellular biomass. In: *Proceedings of the Workshop and EC Contractors Meeting in Capri* (Palz, W. and Pirrwitz, D., Eds.). Reidel Publ. Co., Dordrecht, pp. 184–193.

Gudin, C., Therpenier, C., (1986). Bioconversion of solar energy into organic chemicals by microalgae. Adv Biotechnol Processes. 6, pp. 73–110.

Gustafson KR, Cardellina JH, Fuller RW, Weislo OS, Kiser RF, Snader KM, Patterson GML, Boyd MR (1989). AIDS antiviral sulfolipids from cyanobacteria (blue green algae). Journal of National Cancer Institute, 81, 1254–1258

Hallmann A, Sumper M. The Chlorella hexose/H+ symporter is a useful selectable marker and biochemical reagent when expressed in *Volvox. Proc Natl Acad Sci* USA 93: 669–673, 1996.

Harker M, Bramley PM. (1999) Expression of prokaryotic 1-deoxy-Dxylose- 5-phosphatase in *Escherichia coli* increases carotenoid and ubiquinone biosynthesis. FEBS Lett 448: 115-119.

Harker M, Hirschberg J. (1997). Biosynthesis of ketocarotenoids in transgenic cyanobacteria expressing the algal gene for b-C-4-oxygenase, crtO. *FEBS letters* 404: 129–134.

Harvey (1988). Cracking Open Marine Algae's Biological Treasure Chest. Biotechnology, 6, 488–492.

Hata, J., Hua, Q., Yang., Shimizu, K.M., Taya, M,. (2000). Characterization of energy conversion based on metabolic flux analysis in mixo-trophic liverwort cells *Marchantia polymorpha.* Biochem. Eng. J. 6, pp. 65–74.

Heasman, M., Diemar, J., O' Connor, W., Sushames, T., Foulkes, L., Nell, J.A. (2000). Development of extended shelf-life microalgae concentrate diets harvested by centrifugation for bivalve molluscs—a summary. Aquacult Res. 31 (8 – 9), pp. 637– 59.

Hedenskog, G., Enebo, L., Vendlova, J., Prokes, B.T.I., (1969). Investigation of some methods for increasing the digestibility in vitro of microalgae. Biotechnol Bioeng. 11, pp. 37–51.

Hee-Mock, O., Lee, S.J., Park, M.H., Kim, H.S., Kim, H.C., Yoon, J.H., *et al.* (2001). Harvesting of *Chlorella vulgaris* using a bioflocculant from *Paenibacillus* sp. AM49. Biotechnol Lett. 23, pp. 1229–34.

Hirooka, T., Akiyama, Y., Tsuji, N., Nakamura, T., Nagase, H., Hirata, K., Miyamoto, K., (2003). "Removal of hazardous phenols by microalgae under photoautotrophic conditions". *Journal of Bioscience and Bioengineering,* 95: 200–203.

Holan, Z.R. and Volesky, B. (1994). "Biosorption of lead and nickel by biomass of marine algae". *Biotechnology and Bioengineering,* 43 (11): 1001–1009.

Holan, Z.R., Volesky, B. and Prasetyo, I. (1993). "Biosorption of cadmium by biomass of marine algae". *Biotechnology and Bioengineering,* 41 (8): 819–825.

Hu, Q., Faiman, D., Richmond, A., (1998). Optimal Tilt Angles of Enclosed Reactors for Growing Photoautotrophic Microorganisms Outdoors. J. of Ferment and Bioeng. 85,(2) pp. 230–236.

Hu, Q., Richmond, A,. (1994). Optimizing the population density in *Isochrysis galbana* grown outdoors in a glass column photobioreactor. J. Appl. Phycol. 6, pp. 391–396.

Hu, Q., Richmond, A,. (1996). Productivity and photosynthetic efficiency of *Spirulina platensis* as affected by light intensity, cell density and rate of mixing in a flat plate photobioreactor. J. Appl. Phycol. 8, pp. 139–145.

Hutchins, R.A. (1973). "New methods simplified design of activated carbon systems". *Chemical Engineering Science,* 80: 133–138.

Ibanez Gonzalez, M.J. Robles Medina, A., Molina, G.E., Gimenez Gimenez, A., Carstens, M., Esteban Cerdan, L., (1998). Optimization of fatty acid extraction from *Phaeodactylum tricornutum* UTEX 640 biomass. J. Am Oil Chem Soc. 75, pp. 1735–40.

J Chem Eng 1998; 15: 157–63.

Jaag, O. (1972). "Water quality and water pollution control in Switzerland". *Biological Conservation,* 4(5): 345–354.

Jarvis EE, Brown LM. (1991). Transient expression of firefly luciferase in protoplasts of the green alga Chlorella ellipsoidea. Curr Genet 19: 317–321.

Jeffers, T.H., Ferguson, C.R., Bennet, P. (1991). In *Mineral Bioprocessing,* Smith, R.W. and Misra, M., (Eds.), pp. 289–299. The Minerals, Metals, and Materials Society, Washington, DC, USA. cf. Kratochvil and Volesky (1998).

John, E.H. and Flynn, K.J., (2000). "Modeling phosphate transport and assimilation in microalgae; how much complexity is warranted?", Ecological Modeling, 125, 145–157.

Joshi, J.B., Ranade, V.V., Gharat, S.D., Lele, S.S., 1990. Sparged Loop Reactors. Can. J. Chem. Eng. 68, pp. 705.

Joshi, J.B., Sharma, M.M., (1977). Mass transfer and Hydrodynamic characteristics of Gas inducing type of Agitated contactors. Can. J. Chem. Eng. 65, pp. 683-695.

Kaplan, D., Christiaen, D., Arad, S.M. (1987). "Chelating properties of extracellular polysaccharides from Chlorella spp". *Applied Environmental Microbiology,* 53: 2953–2956.

Kaplan, D., Heimer, Y.M., Abeliovich, A., Goldsbrough, P.B. (1995). "Cadmium toxicity and resistance in Chlorella sp". *Plant Science,* 109: 129–137.

Kapoor A, Viraraghavan T., (1995). Fungi biosorption—an alternative treatment option for heavy metal bearing wastewaters: a review. Bioresour Technol; 53:195–206.

Kapoor A,Viraraghavan T. (1997). Fungi as biosorption. In:Wase DAJ, ForsterCF, editors. Biosorbents for Metal Ions. London, UK: Taylor & Francis; p. 67–85.

Kawaguchi, K,. (1980). Microalgae production systems in Asia. In: Shelef, G., Soeder, C.J. (Eds.), Algae Biomass Production and Use. Elsevier : North Holland Biomedical Press, Amsterdam, pp. 25–33.

Kaya, V.M., Picard, G., (1996). Stability of chitosan gel as entrapment matrix of viable *Scenedesmus bicellularis* cells immobilized on screens for tertiary treatment of wastewater. Bioresource Technol. 56, pp. 147–55.

Kindle KL. High-frequency nuclear transformation of *Chlamydomonas reinhardtii.* Proc Natl Acad Sci USA 87: 1228–32, 1990.

Ko, D. C. K., Porter, J. F., McKay, G. (2000). "Optimized correlations for the fixed bed adsorption of metal ions on bone char". *Chemical Engineering Science,* 55: 5819–5829.

Kobayashi, M., T. Kakizono, K., Yamaguchi, Nishio, N., Nagai, S., (1992). Growth and astaxanthin formation of *Haematococcus pluvialis* in heterotrophic and mixo-trophic conditions. J. Ferment and Bioeng. 74, pp. 17–20.

Kratochvil, D. and Volesky, B. (1998). "Biosorption of Cu from ferruginous wastewater by algal biomass". *Water Research,* 32 (9): 2760–2768.

Kratochvil, D., Volesky, B., Demopoulos, G. (1997). "Optimizing Cu removal/recovery in a biosorption column". *Water Research,* 31 (9): 2327–2339.

Kshirsagar, H.H., Revenkar, M.S., Kamat, M.Y., Lele, S.S. (2002). Cultivation of *Spirulina* in Gas Induced Photobioreactor and isolation of Phycobiliproteins. Ind J.of Biotechnol. 1, pp. 255-262.

Kurtzman AM, Cheney DP. (1991). Direct gene transfer and transient expression in a marine red alga using the biolistic method. *J Phycol* 27 (suppl.): 42.

Laliberté, G., Proulx, G., Pauw, N., De la Noüe, J. (1994). "Algal technology in wastewater treatment". *Ergenisse Limnology,* 42: 283–302.

Langmuir, I. (1918). "The adsorption of gases on plane surfaces of glass, mica and platinum". *Journal of American Chemical Society,* 40: 1361–1403. cf. Kratochvil and Volesky (1998).

Lebeau T, Robert, JM. (2003). Diatom cultivation and biotechnologically relevant products. Part I: cultivation at various scales. *Appl Microbiol Biotechnol* 60: 612–623, 2003.

Lee, C.G., Palsson, B.O. (1994). High-density algal photobioreactors using light-emiting diodes. Biotechnol. Bioeng. 44, pp. 1161–7.

Lee, Y.K., Pirt, S.J., (1981). Energitics of photosynthetic algal growth: influence of intermittent illumination in short (40s) cycles. J. Gen. Microbiol. 124, pp. 43–52.

Lee, Y.K. (1997). Commercial production of microalgae in the Asia Pacific rim J. Appl. Phycol. 9, pp. 403–411.

Lee, Y.K., Ding, S.Y., Low, C.S., Chang, Y.C., Forday, W.L., Chew, P.C. (1995). Design and performance of an *á*-type tubular photobioreactor for mass cultivation of microalgae. J. Appl. Phycol. 7, pp. 47–51.

Lee, Y.K., Low, C.S., (1991). Effect of photobioreactor inclination on the biomass productivity of an outdoor algal culture. Biotechnol. Bioeng. 38, pp. 995–1000.

Li, D.M., & Qi, Y.Z., (1997). Spirulina industry in China: present status and future prospects. J. Appl. Phycol. 9, pp 25–8.

Li,J., Xu,N.S. and Su, W.W., (2003). "Online estimation of stirred tank microalgal photobioreactor cultures based on dissolved oxygen measurements", Biochem. Eng. J., 14, 51–65.

Lima, S.A.C., Castro, P.M.L., Morais, R., (2003). "Biodegradation of p-nitrophenol by microalgae". *Journal of Applied Phycology,* 15: 137–142.

Lin, H.Y., Leu, Y.R., Lan, C.R., Lin, P.P. and Chang, F.L., (2003). "Kinetics of inorganic carbon utilization by microalgal biofilmin flat plate bioreactor", Chemosphere, 53.

Liu RX, Tang HX, Lao WX. (2002b). Advances in biosorption mechanism and equilibrium modeling for heavy metals on biomaterials. Prog Chem; 14: 87–92 [in Chinese].

Lobban, C.S., P.J. Harrison and M.J. Duncan. (1985). The Physiological Ecology of Seaweeds. Cambridge University Press, Cambridge. pp. 209–212.

Luhring H, Witzemann V. (1995). Internodal cells of the giant green alga *Chara* as an expression system for ion channels. *FEMS Lett* 361: 65-69.

Lund H.C. (1995). Freshwater Algae, Their microscopic world explored, Biopress Ltd., Bristol, England.

Luo,H.P., Kemoun, A., Muthanna, H., Al-Dahhan, Sevilla, J. M. F., Sanchez, J.L., Camacho, F.G. and Grima,E.M., (2003). "Analysis of photobioreactors for culturing high value microalgae and cyanobacteria via an advanced diagnostic technique: CARPT., Chemical Engineering Science, 58, 2519–2527, (2003).

Lygren B, Hamre K, Waagbo R (1999). Effects of Dietary Pro and Antioxidants on some Protective Mechanisms and Health Parameters in Atlantic Salmon. Journal of Aquatic Animal Health 11, 211–221.

Macaskie, L.E. and Dean, A.C.R. (1990). In *Biosorption of Heavy Metals,* Volesky, B., (Ed.), (pp. 199–248), CRC Press. cf. Kratochvil and Volesky (1998).

MacKay, D., Salusbury, T., (1988). Choosing between centrifugation and cross flow micro filtration. Chem Eng (Lond). 477, pp 45–50.

Mahajan. G., Kamat, M., (1995). ã-Linolenic acid production from *Spirulina platensis.* Appl. Microbiol. Biotechnol. 43, pp. 466–469.

Mallick, N., (2002). "Biotechnological potential of immobilized algae for wastewater N, P and metal removal: a review". *Biometals,* 15: 377–390.

Marquez, F.J., Sasaki, K., Kakizono, T., Nishio, N., Nagai, S., (1993). Growth characteristics of *Spirulina platensis* in mixo-trophic and heterotrophic conditions. J. Ferment and Bioeng. 76, pp. 408–410.

Martinez Sancho, M.E., Jimenez Castillo, J.M., Espinola Lozano, J.B., El Yousfi, F. (1993). "Sistemas algas–bacterias para tratamiento de residuos liquidos". *Ing. Quim.* 25: 131–135.

McCoy, H.D. (1987). The Commercial Algae; Prospects for One of the Oldest Industries. Aquaculture. Magazine. July/August, 46-54.

McGriff Jr., E.C., McKinney, R.E. (1972). "The removal of nutrients and organics by activated algae". *Water Research,* 6: 1155–1164.

Mendes-Pinto, M.M., Raposo, M.F.J., Bowen, J., Young A.J. Morais, R., (2001). Evaluation of different cell disruption processes on encysted cells of Haematococcus pluvialis: effects on astaxanthin recovery and implications for bio-availability. J Appl Phycol. 13, pp. 19–24.

Metting, F.B. (1996). Biodiversity and application of microalgae. J. Indust. Microbiol. 17, pp. 477–89.

Mezrioui, N., Oudra, B., Oufdou, K., Hassani, L., Loudiki, M., Darley, J., (1994). "Effect of microalgae growing on waste-water batch culture on Escherichia-coli and Vibrio-cholerae survival". *Water Science Technology* 30: 295–302.

Miller PW, Russel BL, Schmidt RR. (1994). Transcription initiation site of NADP-specific glutamate dehydrogenase gene and potential use of its promoter region to express foreign genes in ammonium cultured Chlorella sorokiniana cells. J Appl Phycol 6: 211– 223.

Mohan, F.H. (1978). Improved technologies for the harvesting and processing of microalgae and their impact on production costs. Arch Hydrobiol, Beih Ergeb Limnol. 1, pp 228– 53.

Mohn, F.H. (1980). Experiences and strategies in the recovery of biomass from mass cultures of microalgae. In: Shelef, G., Soeder, C.J.,editors. Algae biomass. Amsterdam: Elsevier; pp. 547– 71.

Molina, G.E., Sanchez Perez, J.A., Garcia Camacho, F., Fernandez Sevilla, J.M., Acien Fernández, F.G., Urda Cardona, J. (1995). Biomass and eicosapentaenoic acid productivities from an outdoor batch culture of *Phaeodactylum tricornutum* UTEX 640 in an air-lift photo bioreactor. Appl. Microbiol. Biotechnol. 42, pp. 658-663.

Mostafa M. El-Sheekh. (2005). Genetic Engineering of Eukaryotic Algae with Special Reference to *Chlamydomonas, Turk journal of Biology,* 29: 65–82.

Munöz, R., Jacinto, M.S.A., Guieysse, B., Mattiasson, B. (2005a). "Combined carbon and nitrogen removal from acetonitrile using algal–bacterial reactors". *Applied Microbiology Biotechnology,* 67 (5): 699–707.

Munöz, R., Köllner, C., Guieysse, B., Mattiasson, B. (2004). "Photosynthetically oxygenated salicylate biodegradation in a continuous stirred tank photobioreactor". *Biotechnology Bioengineering,* 87 (6): 797–803.

Muñoz, R., Rolvering, C., Guieysse, B., Mattiasson, B. (2005). "Photosynthetically oxygenated acetonitrile biodegradation by an algal-bacterial microcosm: a pilot-scale study". *Water Science & Technology,* 51 (12): 261–265.

Munöz, R., Rolvering, C., Guieysse, B., Mattiasson, B. (2005b). "Photosynthetically oxygenated acetonitrile biodegradation by an algal–bacterial microcosm: a pilot scale study". *Water Science and Technology,* 51 (12): 261–265.

Munöz, R., Rolvering, C., Guieysse, B., Mattiasson, B. (2005c). "Aerobic phenanthrene biodegradation in a two-phase partitioning bioreactor". *Water Science and Technology,* 52 (8): 265–271.

Nagels HO, Aach HG, Kreuzaler F. (1989). Transfer and expression of a selectable, chimeric gene in *Chlorella saccharophyla. Braunschweig Symp Appl Plant Mol Biol* pp. 436-441.

Nagle, N.J., Rivard, C.J., Mohagheghi, A., Philippidis, G. (1995). "Bioconversion of cyanide and acetonitrile by a municipalsewage-derived anaerobic consortium". In: Hinchee RE (Eds). Bioremediation of inorganics. Battelle Press, Ohio, pp. 71–79.

Nawaz, M.S., Chaptwala, K.D., Wolfram, J.H. (1989). "Degradation of acetonitrile by Pseudomonas putida". *Applied Environmental Microbiology,* 55: 2267–2274.

Neuhaus G, Neuhaus-Url G, de Groot EJ et al. (1986). High yield and stable transformation of the unicellular green alga *Acetabularia* by microinjection of SV40 DNA and pSV2neo. *EMBO J* 5: 1437–1444.

Nortan L., Kanagaratnam, B., McKenzie T. (2004). "Biosorption of Zn from aqueous solutions using biosolids". *Advances in environmental research,* 8: 629–635.

Nurdogan, Y., Oswald, W.J. (1995). "Enhanced nutrient removal in high rate ponds". *Water Science and Technology,* 31: 33–43.

Onaka, T., Nakanishi, K., Inoue, T., Kubo, S. (1985). "Beer brewing with immobilized yeast". *Biotechnology,* 3: 467–470.

Oswald, W.J. (1988). "Micro-algae and waste-water treatment". In: Borowitzka, M.B.L. (Ed.), Micro-algal Biotechnology. Cambridge University Press, Cambridge, pp. 305–328.

Oswald, W.J. (2003). "My sixty years in applied algology". *Journal of Applied Phycology,* 15: 99–106.

Oswald, W.J., (1988). Large-scale algal culture systems (engineering aspects). In: Pagnanelli F, Esposito A, Veglio F. Multi-metallic modelling for biosorption of binary systems. Water Res 2002; 36: 4095–105.

Palmer, C.M. (1969). "A composite rating of algae tolerating organic pollution". *Journal of Phycology,* 5: 78–82.

Petrusevski, B., Bolier, G., Van Bremen, A.N., Alaerts, G.J., (1995). Tangential flow filtration: a method to concentrate freshwater algae. Water Res. 29, pp. 1419–24.

Phillips Jr, J.N., Myers, J., 1954. Growth rate of Chlorella in flashing light. Plant Physiology. 29, pp. 148–152.

Phushparaj, B., Pelosi, E., Tredici, M., Pinazani, E., Materassi, R., (1997). An integrated system for outdoor production of microalgae and cyanobacteria. J. Appl. Phycol. 9, pp. 113–19.

Pisal, D.S. and Lele, S.S. (2005). Carotenoid production from microalgae, *Dunaliella salina.* Indian journal of Biotechnology, 4, 476–483.

Polli, M., Di Stanislao, M., Bagatin, R., Bakr, E.A., Masi, M., (2002). Bubble size distribution in the sparger region of bubble columns. Chem. Eng. Sci. 57, pp. 197–205.

Porath JB, Gokhman I, Lers A et al. (1997). Developing a transformation system in *Dunaliella. Phycologia* 36 (suppl.): 89.

Qiang, H., Richmond, A. 1994. Optimizing the population density in *Isochrysis galbana* grown outdoors in a glass column photobioreactor. J. Appl. Phycol. 9, pp. 113–19.

Qin S, Sun GO, Jiang P et al. (1999). Review of genetic engineering of *Laminaria japonica* (Laminiriales, Phaeophyta) in China. *Hydrobiologia* 398/399: 469-472.

Ratchford, I.A.J., Fallowfield, H.J., 1992. Performance of a flat plate, air-lift reactor for the growth of high biomass algal culture. J. Appl. Phycol. 4, pp. 1–9.

Richmond, A. (1986). Microalgae of economic potential. In: Richmond, A., editor. CRC handbook of microalgal mass culture. Boca Raton: CRC Press; pp. 199–243.

Richmond, A., Boussiba, S., Vonshak, A., Kopel, R., (1993). A new tubular reactor for mass production of microalgae outdoors. J. Appl. Phycol. 5, pp. 327–332.

Richmond, A., Lichtenberg, E., Stahl, B., Vonshak, A., (1990). Quantitative assessment of the major limitations on productivity of *Spirulina platensis* in open raceways. J. Appl. Phycol. 2, pp. 195–206.

Robinson, P.K. (1998). "Immobilized algal technology for wastewater treatment purposes". In: Wong, Y.S., Tamnora, F.Y. (Eds.), Wastewater Treatment with Algae. Springer and Landes Bioscience, pp. 1–16.

Rodrýguez-Maroto, J.M., Carlos, J., Jose A., Niell, F.X., (2005). Air bubbling results in carbon loss during microalgal cultivation in bicarbonate-enriched media: experimental data and process modeling. Aquacult Eng. 32, pp. 493–508.

Rose, P.D., Boshoff, G.A., van Hille, R.P., Wallace, L.C.M., Dunn, K.M., Duncan, J.R. (1998). "An integrated algal sulphate reducing high rate ponding process for the treatment of acid mine drainage wastewaters". *Biodegradation,* 9: 247–257.

Rossignol, N., Vandanjon, L., Jaouen, P., Quemeneur, F. (1999). Membrane technology for the continuous separation microalgae/culture medium: compared performances of cross-flow micro-filtration and ultrafiltration. Aquacult Eng. 20, pp. 191– 208.

Rossler PG, Bleibaum JL, Thompson GA et al. Characteristics of the gene that encodes acetyl-CoA carboxylase in the diatom *Cyclotella cryptica. Ann N Y Acad Sci* 721: 250–256, 1994.

Roy, D. S., Chet, I., Mitchell, R., 1978. Quantitative Assay for algal Chemotaxis. Appl. and Environ. Microbiol. Dec, pp. 847–850.

Ruane, M., 1977. Extraction of caroteniferous materials from algae. Australian patent No. 7,239, 574.

Safonova, E., Dmitrieva, I.A., Kvitko, K.V. (1999). "The interaction of algae with alcanotrophic bacteria in black oil decomposition". *Resource Conservation Recycling,* 27: 193–201.

Safonova, E., Kvitko, K.V., Iankevitch, M.I., Surgko, L.F., Afri, I.A., Reisser, W. (2004). "Biotreatment of industrial wastewater by selected algal–bacterial consortia". *Engineering Life Science,* 4: 347–353.

Sanchez, M.A., Camacho, F.G., Gomez, A.C., Molina, G.E., Chisti,. Y. (2000). Bubble-Column and Airlift Photobioreactors for Algal Culture, AIChE J. 46 (9), 1872–1887.

Sanchez, M.A., Garcýa, M.C.C., Camacho, F.G., Molina, G.E., Chisti, Y., (2002). Growth and biochemical characterization of microalgal biomass produced in bubble column and airlift photobioreactors: studies in fed-batch culture. Enzyme and Microbial Technol. 31, pp. 1015–1023.

Santos CA, Vieira AM, Fernandes HL, Empis JA, Novais JM (2001). Optmization of the biological treatment of hypersaline wastewater from *Dunaliella salina* Journal of Chemical Technology and Biotechnology, 76, 1147–1153.

Sato T., Usui S., Tsuchiya Y. and Kondo Y., 'Invention of outdoor closed type photobioreactor for microalgae' Energy Conversion and Management, Volume 47, Issue 6, April 2006, pp. 791–799.

Schafer, R., Mertern, C., Eigenberger, G. (2002). Bubble size distributions in a bubble column reactor under industrial conditions. Exp Thermal Fluid Sci. 26, pp. 595–604.

Schiewer S, Volesky B. 1997. Ionic strength and electrostatic effects in biosorption of divalent metal ions and protons. Environ Sci Technol; 1: 2478–85.

Schiewer S, Volesky B. (1996). Modeling multi-metal ion exchange in biosorption. Environ. Sci. Technol.; 30: 2921–27.

Schiewer S, VoleskyB. (1995a). Mathematical evaluation of the experimental and modeling errors in biosoprtion. Biotechnol Technol; 9:843–48.

Schumacher, G., Blume, T., Sekoulov, I. (2003). "Bacteria reduction and nutrient removal in small wastewater treatment plants by an algal biofilm". *Water Science and Technology,* 47: 195–202.

Semple, K.T., Cain, R.B., Schmidt, S. (1999). "Biodegradation of aromatic compounds by microalgae". *FEMS Microbiology Letters,* 170: 291–300.

Shamala, T.R., Drawert, F., Leupold, G., Becker, W. (1982). Studies on *Scenedesmus acutus* growth. The incorporation of 14C-glucose by *Scenedesmus acutus* under light and dark periods. Biotechnol. and Bioeng. 24 (5), pp. 1183–1190.

Shanbhag G (2001). Bioactive Compounds from Microalgae, A Ph. D. thesis submitted to University of Mumbai.

Sharma, D.C. and Forster, C.F. (1993). "Removal of hexavalent chromium using sphagnum moss peat". *Water Research,* 27 (7): 1201–1208.

Siegel, B.Z., Siegel, S.M. (1973). *CRC Crit. Rev. Microbiol.* 1973, 10. cf. Crist *et al.* (1981).

Silva, H.J., Cortinas, T., Ertola, R.J. (1987). Effect of Hydrodynamic Stress on *Dunaliella* Growth. J. Chem. Technol. Biotechnol. 40, pp. 41.

Singh, R., and Prasad, B.B. (2000). "Trace metal analysis: selective sample (copper II) enrichment on an AlgaSORB column". *Process Biochemistry,* 35: 897--905.

Skulberg Olav (2000). Microalgae as a source of bioactive molecules – experience from cyanophyte research. Journal of Applied Phycology 12, 341–348.

Sommer, T., (1988). Commercial Microalgal Production: The State of the Art. Aqu. Mag. March/April. pp. 44–47.

Sontaya, K., Loataweesup, W., Powtongsook, S., Prasert P. (2005). Optimal growth conditions and the cultivation of *Chaetoceros calcitrans* in airlift photobioreactor. Chem. Eng. J. 105, pp. 91–98.

Specktorova, L., Le Roy, Creswell, R., Vaughan, D. (1997). Closed tubular cultivations: an innovative system for commercial culture of microalgae. World Aquacult. 28, pp. 39–43.

Subaramaniana, G., Uma, L. (1997). "Role of cyanobacteria in pollution abatement". In: Sinha, M.P. (Ed.), Recent Advances in Ecobiological Research, Vol. 1. APH Publishing Corporation, pp. 435–443.

Suh JH, KimDS. (2000). Effects of Hg^{2+} and cell conditions on Pb^{2+} accumulation by Saccharomyces cerevisae. Bioprocess Eng. 23: 327–29.

Suzuki, T., Matsuo, T., Ohtaguchi, K., Koide, K. 1995. Gas Sparged Bioreactors for CO_2 Fixation by *Dunaliella tertiolecta*. J. Chem. Technol.Biotechnol. 62, p. 351.

Torzillo, G., Carlozzi, P., Pushparaj, B., Montaini, E., Materassi, R. (1993). A two-plane tubular photobioreactor for outdoor culture of *Spirulina*. Biotechnol Bioeng. 42, pp. 891–898.

Torzillo, G., Pushparaj, B., Bocci, F., Balloni, W., Materassi, R., Florenzano, G. (1986). Production of *Spirulina* biomass in closed photobioreactors. Biomass. 11, pp. 61–74.

Travieso, L., Benitez, F., Weiland, P., Sanchez, E., Dupeyron, R., Dominguez, A.R. (1996). Experiments on immobilization of microalgae for nutrient removal in wastewater treatments". *Bioresource Technology,* 55: 181–186.

Tredici, M.R, Chini Zittelli, G. (1997). Cultivation of *Spirulina* (Arthrospira) *patensis* in flat plate reactors. In:Vonshak A, editor. *Spirulina platensis* (Arthrospira): physiology, cell-biology and biotechnology. London: Talyor and Francis, pp. 117–30.

Tseng, C.K. (1981). Commercial Cultivation. In The Biology of Seaweeds. C. S. Lobban and M.J. Wynne eds. Blackwell Scientific Publication, Oxford.

Tsukada, O., Kawahara, T., Miyachi, S. (1977). Mass culture of *Chlorella* in Asian countries. In: Mitsui, A., Miyachi, S., San Pietro, A., Tamura, S. (Eds.), Biological Solar Energy Conversion. Academic Press, New York, pp. 363–365.

Van Hille, R.P., Boshoff, G.A., Rose, P.D., Duncan, J.R. (1999). "A continuous process for the biological treatment of heavy metal contaminated acid mine water". *Resources Conservation and Recycling,* 27: 157–167.

Venkataraman, Bhagyalakshmi, Ravishankar (1995). Commercial potentials of micro and macro algae-problems and potentials. Indian Journal of Microbiology, Volume 35, Number 1.

Venkataraman, L.V., (1978). New possibility for microalgae production and utilisation in India. Arch Hydrobiol Beih .11, pp. 199–210.

Vijayaraghavan, K., Jegan, J., Palanivelu, K., Velan, M. (2005). "Biosorption of copper, cobalt and nickel by marine green alga Ulva reticulata in a packed column". *Chemosphere,* 60: 419–426.

Vijayaraghavan, K., Prabu, D. (2006). "Potential of *Sargassum wightii* biomass for copper (II) removal from aqueous solutions: Application of different mathematical models to batch and continuous biosorption data". *Journal of Hazardous Materials,* Article in press.

Volesky B. and Holan Z.R. (1995). "Biosorption of heavy metals" *Biotechnology progress* 11: 235–250.

Volesky B. (1993). Cadmium biosorption by Saccharomyces cerevisiae. Biotechnol. Bioeng.; 41: 826–29.

Volesky, B. and Votruba, J. "Modeling and optimization of fermentation processes", Elseveir Science Publishers, Amsterdam, Vol. 1., (1992).

Vollenweider, R.A. (1985). "Elemental and biochemical-composition of plankton biomass—some comments and explorations". *Arch. Hydrobiol,* 105: 11–29.

Vonshak Avigad (1990). Recent advances in Microalgal Biotechnology, Biotech Advances 8, 709-727.

Vonshak, A. [Ed.]., (1997). *Spirulina platensis* (Arthrospira): Physiology, Cell-Biology and Biotechnology. Taylor & Francis, London, p. 233.

Waaland, J.R. (1981). Commercial Utilization. In The Biology of Seaweeds. C.S. Lobban and M.J. Wynne eds. Blackwell Scientific Publication, Oxford.

Wang CW, Oh MK, Liao JC. (1999). Engineered isoprenoid pathway enhances astaxanthin production in Escherichia coli. Biotechnol. Bioeng. 62: 235–241.

Wang JL. (2002). Immobilization techniques for biocatalysts and water pollution control. Beijing: Science Press; [in Chinese].

Watanabe, Y., de la Nou¨e, J., Hall, D.O., (1995). Photosynthetic performance of a helical tubular photobioreactor incorporating the cyanobacterium *Spirulina platensis.* Biotechnol. Bioeng. 47, pp. 261–269.

Website www.bccresearch.com as retrieved on 5th March 2007.

Website www.siu.edu as retrieved on 5th March 2007.

Weiland, P., Onken, U., (1981). Differences in the behaviour of bubble columns and airlift loop reactors. Ger. Chem. Eng. 4, pp. 174–181.

Weissman, J.C., Goebel, R.P., Benemann, J.R. (1988). Photobioreactor design: Mixing, carbon utilization, and oxygen accumulation. Biotechnol. Bioeng. 31, pp. 336–344.

Weller C, Lustbader J, Seshadri K, Brown JM, Chadwik CA, Kolthloff CE, Ramnarain S, Pollak S, Canfield R, Homans SW (1996). Structural and Conformational Analysis of glycan moeties insitu on isotopically ^{13}C, ^{15}N-enriched recombinant human chorionic gonadotropin. Biochemistry, 35, 8815–8823.

Wilde, E.W., Benemann, J.R. (1993). "Bioremoval of heavy metals by the use of microalgae". *Biotechnology Advances,* 11: 781–812.

Wu, J., Goosen, M.F.A. (1995). Evalutation of the killing volume of gas bubbles in sparged animal cell culture bioreactors. Enzyme Microb Technol. 17, pp. 1036–1042.

Xiaoxi, W., Merchuk, J.C. (2004). Simulation of algae growth in a bench scale internal loop airlift reactor. Chem. Eng. Sci. 59, pp. 2899–2912.

Yang JB, Volesky B. (1999). Modelling uranium-proton ion exchange in biosorption. Environ. Sci. Technol. 33: 4079–85.

Yu, R.Q., Wang, W-X. (2004). "Biokinetics of cadmium, selenium, and zinc in freshwater alga Scenedesmus obliquus under different phosphorus and nitrogen conditions and metal transfer to Daphnia magna". *Environtal Pollutants,* 129: 443–456.

Yuan A, Knauf V. (1997). Modification of plant components. Current Opinions in Biotechnology, 8, 227–233.

Zaslavskaia LA, Lippmeier JC, Kroth PG et al. (2000). Transformation of the diatom *Phaeodactylum tricornutum* (Bacillariophyta) with a variety of selectable marker and reporter genes. *J Phycol* 36: 379–386.

Zoha S, Ramnarain S, Allnutt FCT (1998). Ultrasensitive direct fluorescent immunoassay for thyroid stimulating hormone. Clinical Chemistry, 44(9), 2045–2046.

INDEX

A

B

C

D

E

F

S

T

U

V

W

X

Y

Z